Grow *your own* Paper

D0598528

maureen richardson

Martingale
& COMPANY

TM

A QUARTO BOOK

First published in 1999 by Martingale & Company
PO Box 118
Bothell, WA 98041-0118
www.patchwork.com

Copyright © 1999, Quarto Publishing Inc.

All rights reserved. This book may not be reproduced in whole or in part, in any form or by any means, electronic or mechanical, including photocopying and recording, or by any information storage and retrieval system now known or hereafter invented without prior written permission from the copyright holder.

Library of Congress Cataloging-in-Publication data available.

ISBN 1-56477-280-2

QUAR.GYO

Conceived, designed, and produced by
Quarto Publishing plc
The Old Brewery
6 Blundell Street
London
N7 9BH

Editor Anna Watson
Art Editor Sally Bond
Assistant Art Director Penny Cobb
Text Editors Claire Waite, Sam Merrell
Designer Julie Francis
Photographers Colin Bowling, Pat Aithie, Les Weis
Illustrator Peter Campbell
Indexer Diana LeCore

Art Director Moira Clinch
Publisher Piers Spence

Manufactured in China by Regent Publishing Services, Ltd.
Printed in Singapore by Star Standard Industries (pte) Ltd.
04 03 02 01 00 99 6 5 4 3 2 1

Safety Notice

Papermaking can be dangerous, and readers should follow safety procedures and wear protective clothing and goggles at all times during the preparation of chemicals and cooking of alkali pulp mixtures. Neither the author, copyright holders, nor publishers of this book can accept legal liability for any damage or injury sustained as a result of making paper.

Contents

Getting Started

Almost everyone has some experience of gardening, even if it is restricted to a patio, a window box, or even a houseplant or two. Whether the purpose is to admire them, smell them, or eat them, growing plants gives us great pleasure, a pleasure which can be further increased when another incentive is added—the use of plants to make beautiful papers.

Making natural papers

Papermaking itself is a gratifying occupation. It is easy to get into and splendid results are quickly achieved. No previous experience is necessary—you can start right away and be pleased with what you produce. As time goes on, you will gain the skill and confidence to become more adventurous in your techniques, and more creative in your invention. Hand papermaking does not demand much monetary outlay. Little special equipment is needed, and you will find that almost all the tools and materials are already on hand in your kitchen or utility room.

The most gratifying aspect of hand papermaking, however, is its positive effect on the environment. We consume a great deal of paper, putting harmful pressure on the environment, so it is wise to recycle as much of it as we can. Papermaking by hand actively recycles, and the cultivation of the fresh plant ingredients has an almost entirely benign ecological impact: as with food production, the idea that you can "grow your own" paper is a sound principle to follow.

Crafting with handmade paper

Having made it, what will you do with it? Your hand-crafted plant papers are special and not intended to replace the papers we use every day. You may want to send special messages on personally crafted paper, such as seasonal greetings, love letters, and even poems. You can create presents with it, cover books and boxes, make calling cards, mount photographs and craft pieces on it, embroider it, or wrap jewels in it. You can even print images on it—recycled paper blended with plant material takes ink beautifully from linocuts and wood blocks.

Some of the various uses I have found for handmade paper are illustrated on the next pages, but you will soon be able to express yourself in a unique way that is completely personal, through the medium of paper. I am confident that you will come to love making your own papers as much as I do.

Handmade Papers

Shown here are a few examples to suggest to you how you might use your own papers. You can stitch, glue, or fold them for a great variety of purposes—screens, endpapers, bookmarks, business cards, baskets, decorated picture frames, and so on . . .

If you wish to use the paper for writing, you must either add size into the vat before you form the sheets (see p. 98) or brush a coat of white craft glue (P.V.A.) over the dry sheet.

Yucca lampshade
Paper made from pure yucca pulp has the right degree of translucency and a warm colo for lampshades.

Valentine's card
This Valentine's card, in mixed flower paper, is diplomatically concealed in a pure white envelope.

Writing paper folio
To make this folio, I bonded handmade paper to scored card. It contains sized envelopes and sheets of paper for writing on.

Book covers
These store-bought sketch books are personalised with marigold and mixed flower papers.

Paper wallet
I keep samples of my papers in a stitched wallet to show to clients.

Floral collage
The artist Hilda Robinson has used one of my handmade papers as a background for a framed collage of pressed flowers.

Gift boxes
These tiny boxes are covered with pressed flower papers and corn silk paper.

Plant Materials

To make your own plant papers you must first find the materials from which they are made. They can be raised in your garden, found growing wild, turn up as by-products from your kitchen, or be bought from stores.

Growing

Annuals and biennials Annual plants—which develop from seed, flower, and produce seeds in one year—do not make great demands on the soil; any well-drained soil will do. The same applies to biennials, which are sown in one year and go on to flower and set seed the next. You do not need to prepare the ground to any great depth, nor add nutrients—enriched soil encourages them to produce leaves rather than flowers.

I sow hardy annuals—such as love-in-a-mist, cornflower, marigold, violas, and grasses—in the fall, to provide early cut flowers the next year. I also keep cutting the flowers to encourage further production.

I sow seed for biennials—such as salsify, honesty, and wallflower—in the spring and early summer for the next year's crop. To produce parsley leaf, however, the seed needs to be sown twice; in spring and summer.

Perennials Herbaceous perennials—like hydrangea, larkspur, fennel, and yucca—flower and seed year after year, and are easy to grow. They need a reasonable depth of fertile, well-drained loam, so I cultivate them fairly deeply and add garden compost.

Trees and shrubs I give minimal attention to the cultivation of trees and shrubs, watering in times of drought, and controlling their size by pruning.

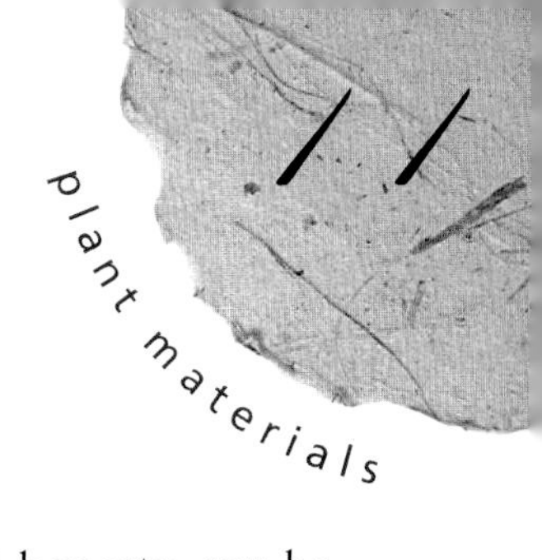

Gathering and Storing

Flowers In the flower garden I take a little time to enjoy the blooms on the plant before harvesting them for paper. I rescue fading cut flowers from the vase rather than consigning them to the compost heap. Some flowers may be used fresh in paper, but I often add some pressed specimens, which retain their distinctive shape in the paper.

Drying, Pressing, and Storing Pick flower heads on a dry morning. Pull the petals off flowers with hard centers or with heads made up of florets. Space them on a sheet of blotting paper or kitchen towel, cover with another sheet, and insert between the pages of a telephone directory. Put a weight on top and check after a week. If the flowers are dry, store them in a dark, dry place. Petals of tulips, roses, and larkspur dry best on a muslin-covered picture frame in a dark, airy position. Turn them daily until dry, and store as above.

Seeds Some plants, like salsify and honesty, can be enjoyed to their full extent in the garden, until they set seed. Collect the seed heads and keep them in a dry jug until you are ready to use them.

When walking in the countryside I often see valuable seed material, such as old-man's-beard, wild parsnip, thistledown, and cattail spikes. These can be gathered and hung upside down in loosely tied paper or net bags until they dry and shed their seeds.

Other Parts of Plants The garden yields plenty of plant material other than flowers and seeds. From my garden I have collected yucca leaves, the lichen-like oakmoss from an old apple tree, skeleton leaves found underfoot, the tendrils of ivy, willow prunings, and maple leaves, all gathered as I notice their potential and kept until needed. I have also retrieved pondweed from a river nearby, which appears plentifully but briefly in the summer.

Begging, Buying, and Reclaiming Many plants grown for food have parts that can be put aside for papermaking, like the silks from corn cobs, the outer leaves of leeks, and banana skins.

In other cases the necessary material may have to be bought from stores. I use teas bought from the grocer and buy the dried roots of dye plants such as alkanet, logwood, and madder, sisal fiber, and raffia from specialist craft stores. Linseed, parsnip, and salsify seeds from a nursery or garden center.

Plant scents

As you will see from the recipes, some of the plants naturally bring their own scent to the papers you make. This will not always last very long, so you may wish to add a ready-distilled essential oil to the pulp. These are readily available from toiletry or New Age healthcare stores.

Pulp Materials

Most recipes in this book are based on a basic pulp made from recycled paper. Industrially produced paper uses such large quantities of virgin wood pulp that the world's tree cover is drastically threatened. To help remedy this it is essential that we recycle wastepaper. But paper cannot be recycled indefinitely; the repulping process makes the fibers shorter and shorter until they are reduced to dust and incapable of bonding together.

Wastepaper

Fortunately for hand papermakers, it is easy to obtain high quality wastepaper and produce excellent pulp from it. Computers and photocopiers inevitably produce wastepaper, which generally is of a very good quality. Some computer paper has green lines on it, and the print turns purple as it dissolves, resulting in a pale blue pulp—not unattractive if that is what you want.

Wastepaper with a lot of print causes greyness in the pulp, so select blank sheets if you require a completely pure white result. Conversely, to obtain a tint in your paper, add colored card or paper napkins to the pulp.

Scraps of acid-free mat or mount board from picture framers make excellent paper. On a similar note, spent sheets of artists' watercolor paper or Bristol board (cartridge paper) also produce wonderful paper.

Newsprint

I never make pulp from newsprint. It is of the lowest quality and has already been recycled to the limit. It is not acid-free, so will go yellow and brittle with age.

Linters

From craft shops and art supply stores, you can buy ready-processed pulp material. It is available in board form, needing only hydrating, or as custom beaten pulp in containers. It is made from cotton which is a readily available fiber, and in this form it is known as "cotton linters."

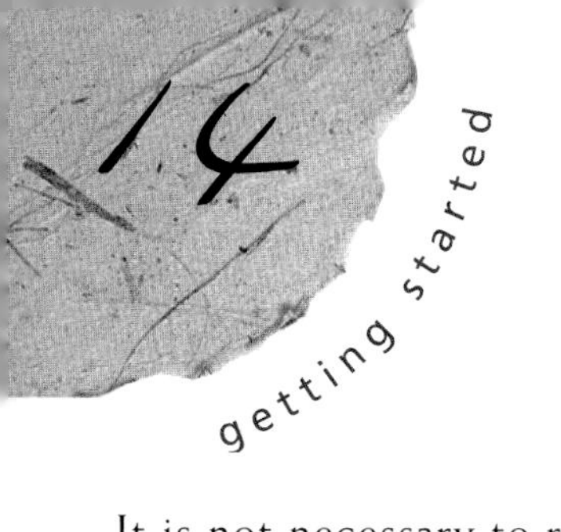

The Workshop

It is not necessary to rush out and hire a workshop for your papermaking, since you can probably use your existing domestic space with some minor adaptations, particularly if you are lucky enough to have a large kitchen or a utility room.

The space you will need can be broadly divided into "wet" and "dry" areas. In the wet area you need a hot and cold water supply and drainage, and an electric socket (see p. 16 for a note on using electricity near water). This is where you will make the pulp, process fibers, work in the vat, form sheets, and carry out the initial drying. Your kitchen may well be suitable.

The dry area can be part of a large kitchen or a separate room. Here the newly made sheets are dried, processed, and stored. Your dry source material and any chemicals will also be kept here.

Some of the "adventurous" recipes involve cooking with caustic solutions. This should be done outside, in a space that is sheltered enough to prevent papers blowing around but open enough to disperse the vapours. Weather permitting, it can be pleasant to do most of the paper-making processes outside, such as work at the vat and pressing and drying, but sheets should be air-dried away from direct sunlight.

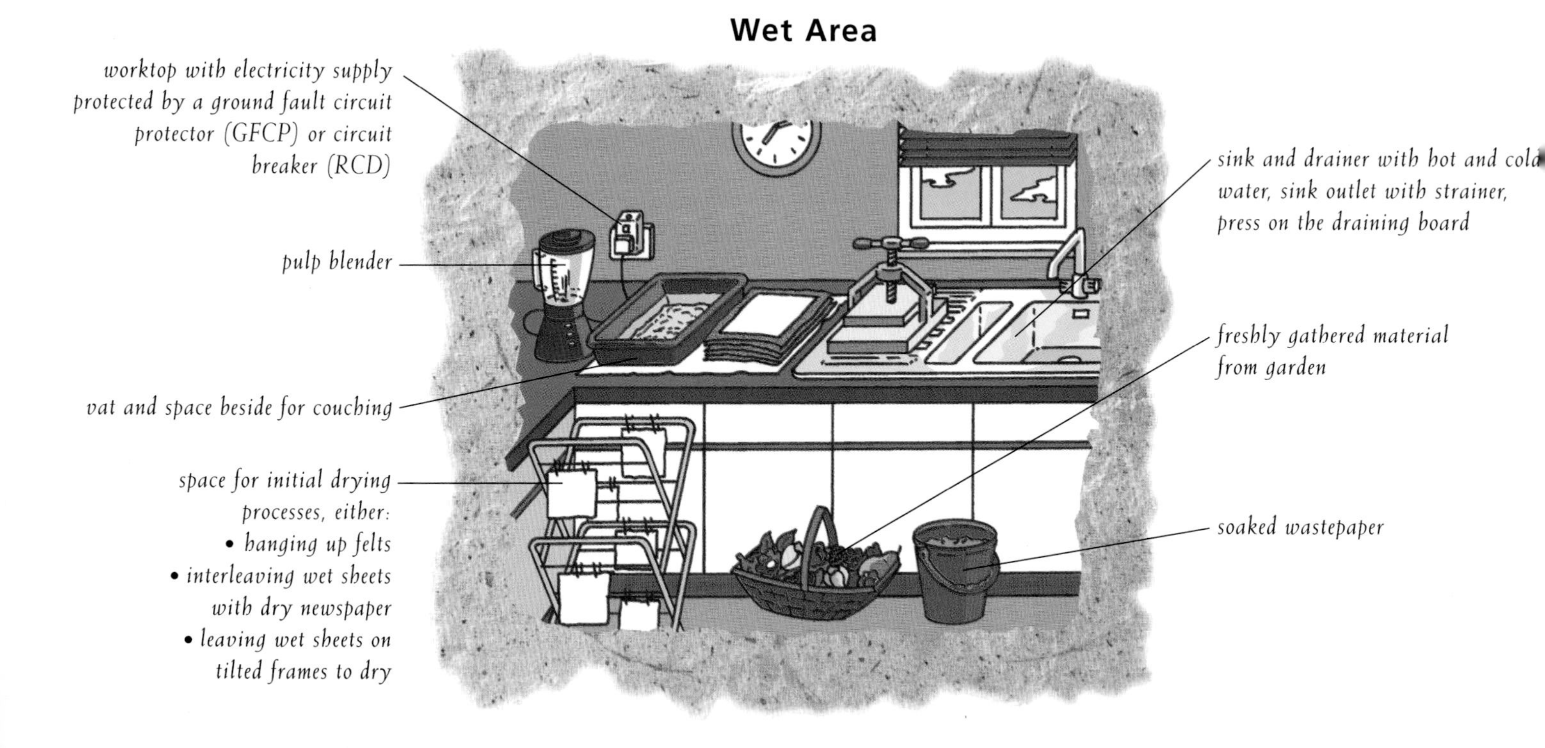

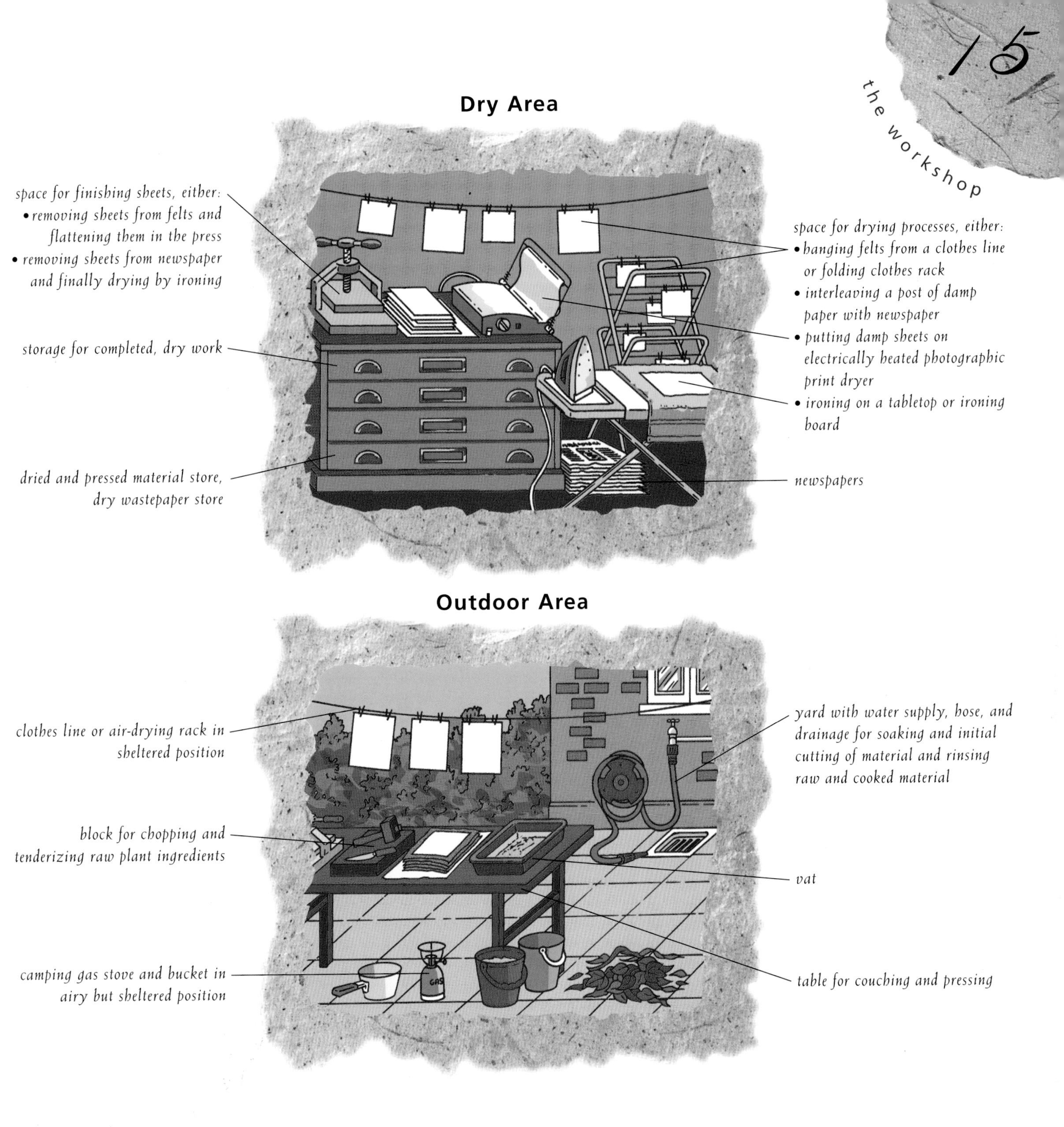
Dry Area
space for finishing sheets, either:
• removing sheets from felts and flattening them in the press
• removing sheets from newspaper and finally drying by ironing
storage for completed, dry work
dried and pressed material store, dry wastepaper store
space for drying processes, either:
• hanging felts from a clothes line or folding clothes rack
• interleaving a post of damp paper with newspaper
• putting damp sheets on electrically heated photographic print dryer
• ironing on a tabletop or ironing board
newspapers
Outdoor Area
clothes line or air-drying rack in sheltered position
block for chopping and tenderizing raw plant ingredients
camping gas stove and bucket in airy but sheltered position
yard with water supply, hose, and drainage for soaking and initial cutting of material and rinsing raw and cooked material
vat
table for couching and pressing
GAS

Papermaking Equipment

Basic Papermaking Equipment

A Blender You may already have a blender or liquidizer that is suitable for mixing the pulp. It must have a 2-pint (1.2-l) capacity jar. If you are buying one specifically for papermaking, get one with a coffee grinder attachment, very useful for reducing hard, dry material like dye chips.

Safety box relating to the blender
Any electrical appliance which is to be used in a wet area or outside should be fitted with a circuit breaker—a ground fault circuit protector (GFCP), also known as a residual current device (RCD). Such a device may already be fitted within your whole circuit. If not you can buy an adapter that is plugged into the socket.

The Vat The vat is the open-topped vessel containing the pulp while sheet forming. Its size depends on the largest mold you are using: you need to have enough room in the vat to keep hold of each side of the mold while you are scooping it below the surface of the pulp.

A standard, rectangular dishwashing pan or bowl makes a good starting vat

for making small sheets. As they are inexpensive it is a good idea to have several. You could also use a cat-litter tray, which gives a convenient surface area for scooping the pulp. To make large sheets, a vat can be adapted from a baby's bath tub or a fiberglass lily pond.

The Mold This is at the heart of papermaking, since this is what the sheet is formed on. Commercial hand papermakers use the traditional, hardwood, brass-bound mold and deckle with a mesh of phosphor-bronze wire.

On a domestic scale something much simpler is appropriate. To get you started, try ready-made embroidery frames, available as circles, ovals, and squares. Stretch mesh tightly over the frame and secure it in place with the outer ring. Net curtain material (Terylene) can be used as the mesh. Use the plain variety rather than those with raised patterns, which are only useful for embossed papers. You will get differing results depending on

the size of the holes in the net mesh (large holes let the finest fibers escape). Try meshes between 200 and 1200 holes per square inch (6.5cm²); 600 holes is best.

Wire mesh window screen (250 holes) is more robust than curtain net but is not as nice to handle; cover the cut edges to protect your fingers. With plastic window screen you do not have this problem.

You can make rectangular molds from self-assembly stretchers for artists' canvases from art supply stores. The stretcher bars come in pairs with ready-machined bridle joints. They are simply tapped together with a mallet and a smear of waterproof glue in the joints.

When buying frames remember that the size quoted for the artist is the outer frame size, and the inside area, which is the size your sheet will be, is smaller. To secure the mesh use a staple gun with stainless steel staples or use brass drawing pins.

An option is to use rigid plastic canvas as a mold, which at sizes up to 6 x 5 in (15 x 12.5cm) is rigid enough not to need a frame and can be cut into shapes for collage designs (see p. 107). Another frameless option is to use split-bamboo placemats or sunbathing mats (see p. 109).

The Felts The wet sheets of paper are interleaved between felts during one of the drying processes (see p. 102). Traditionally felts were made of wool, giving watercolor paper its texture. For domestic papermaking, felts of man-made materials have many advantages. They are easily rinsed and inexpensive, so can be disposed of if they become stained.

Striped viscose dishcloths are ideal as inexpensive and readily available felts. Pellon (vilene), sold as an interfacing to dressmakers, is a deluxe choice for felts. The least expensive felts are disposable diapers (nappy liners), but their small size is a limitation.

Water Absorbers Capillary matting is worth tracking down at your nursery or garden center. You stand your houseplants on it for self-watering if you go away on vacation. Cover your work surface with it, and put a layer at the top and bottom of the pile when pressing a stack of wet paper (a post). Cheap alternatives are old towels or newspapers, but these do not compare in convenience.

The Press The press is used to expel water from a post of freshly made sheets, and a specialist screw press is ideal, but not essential. A pair of melamine-faced plywood boards and four G clamps (or the equivalent ratchet type clamps or solo clamps) will serve very well.

Drying Apparatus A clothes line hung across part of your work area can be used for drying the sheets, but a folding clothes rack is more compact. Use an iron, applied through thick paper, to flatten and dry some sheets.

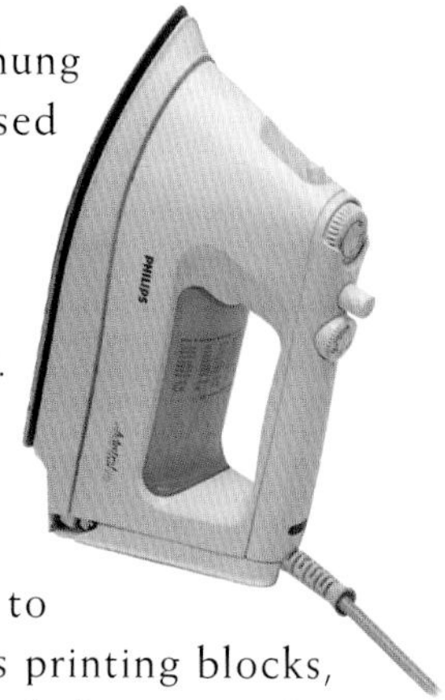

Embossing Equipment Almost anything can be used to emboss paper, such as printing blocks, lace, and textured cloth (see p. 108).

Watermarking Equipment Use fine stainless steel wire and pliers to make a design then sew it onto the mesh with a needle and thread (see p. 108).

Useful Extras For your own protection, a toweling apron, not a plastic one, plus gloves or barrier cream.

A supply of dry newspaper or absorbent paper comes in useful for various purposes. Use fine-mesh netting over your tap to filter out impurities and silt.

For working with the paper pulp, a range of plastic jugs, sieves, and bowls (with shower caps to use as temporary lids) all come in useful.

Scissors, broad and narrow paintbrushes, tweezers, craft glue (P.V.A. adhesive), spray starch, spray mister, and a pointed palette knife all come in useful for adding materials to the paper and getting it off a mold.

For storing prepared flowers and pulp, use freezer boxes. For dried flowers use transparent food bags, labeled with a permanent marker.

To record your progress, use a folder with transparent sleeves to keep paper samples and notes.

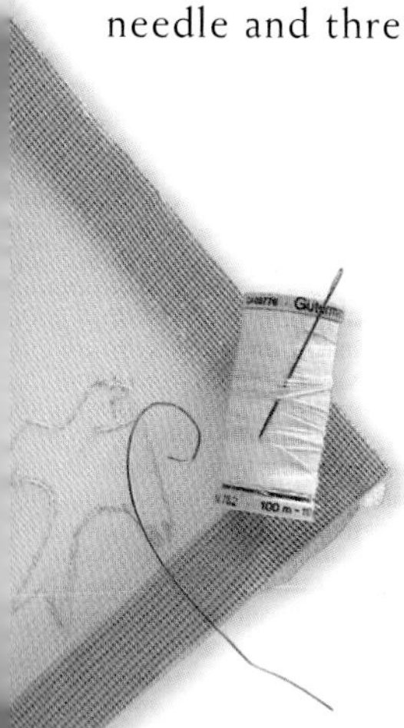

Intermediate Papermaking Equipment

The only extra process involved in the intermediate recipes is some cooking. For this you will need a kettle, a 2-pint (1.2-l) saucepan, and a jam thermometer. No chemicals are involved in these recipes, so you can cook them on your kitchen stove.

Adventurous Papermaking Equipment

As well as the equipment used for basic and intermediate papermaking, the adventurous recipes require a few extras. These recipes involve handling and cooking caustic solutions, which should be done outside using a camping gas cooker. Avoid aluminum containers—use a stainless steel or galvanized bucket or a large stainless steel or unchipped enamel saucepan. For protection from the chemicals, you must wear stout waterproof gloves, a mask, and goggles.

You will also need: wooden tongs; net bags to hold plant materials; a mallet and chopping board; large and small kitchen knives; pH test papers; measuring spoons; and clearly labeled glass jars for holding chemicals.

Equipment for Dedicated Papermakers

If you find you are making a lot of paper you may wish to speed up and ease the process.

When cutting up long-stemmed material into smaller pieces I use an old agricultural chaff cutter. A garden shredder also does the same job.

When preparing lots of pulp in a deep container make use of an electric drill with a paint stirring attachment.

A turkey baster is useful for controlled placing of pulp when you are building up a design on a sheet.

I have an electric ironing machine that quickly and evenly presses and dries several small sheets at a time. Otherwise the drying process can be speeded up by using a photographic print dryer, or by putting a dehumidifier in the room.

Paper Recipes

The main ingredient of paper pulp is plant cellulose, and although paper can be made from any plant, the higher the ratio of usable cellulose in the plant the stronger the pulp and the better the paper. Pulp is a carrier, a fibrous web to which you can add any decorative plant material that your imagination suggests.

This section contains 50 paper "recipes," which add various plant materials to the paper pulp. Some of the more adventurous recipes create a pure plant pulp rather than adding ingredients into a base of paper pulp. These recipes are just a sample of the limitless possibilities available to you, but I have based my selection on readily available ingredients, most of which can be found in the home, grown in the garden, or collected from plants in the wild.

Time and preparation

While the majority of these recipes require very little in the way of special equipment or space, the one thing you will need is time. Papermaking should not be hurried, and you will gain the most enjoyment and satisfaction from ensuring that you can work in peace and quiet, without interruptions. I recommend putting aside three hours of uninterrupted time for papermaking. Above all papermaking is a soothing activity and the more relaxed you are the better the results.

Preparation is also an important aspect of papermaking and you should take time to read and understand the techniques (see pp. 98–109), assemble your equipment, and organize your workspace before starting. Each recipe is marked as either simple, intermediate, or adventurous, indicating how complex and time-consuming it is and the equipment that you will need (see pp. 16–19). For ease of reference, we have divided the recipes into two sections, cooked and uncooked.

Variations and experiments

Although the recipes are illustrated with my own work, the unpredictable nature of papermaking means that your results will differ from mine. This is one of the joys of papermaking—and also one of the frustrations. Even when you share the same pulp and vat with another papermaker your results will never be identical. And remember, new ingredients can be used to make paper too—experiment with whatever grows in your garden to make totally new papers.

Uncooked papers

With these recipes you will be able to start papermaking with minimal preparation, equipment, and space. They are all based on making a pulp from recycled paper, to which plant ingredients such as flowers, leaves, bark, and seeds are added. Instructions for making the basic paper pulp are on page 98, and each batch makes about 12 sheets. If the amount of plant material to add is not specified, simply use what you have until you achieve the look you want.

Viola

Viola tricolor

The viola is also known as heartsease or wild pansy, and Shakespeare said that maidens used to call it "love in idleness." The bloom avoids injury from rain or dew by drooping during storms and at night.

equipment

Basic papermaking equipment (see pp. 16–18).

cultivation

Viola is an annual or a short-lived perennial that will grow in sun or shade in well drained but moisture-retentive soil. Sow seeds in spring in rich, damp soil, for flowers in late summer. Seeds sown in fall will flower in early summer.

harvesting and preparation

Use the flowers only. Pick the flower heads through the summer. Press and dry them (see p. 11).

beating method

No extra beating is required once the pulp is prepared (see p. 98).

sheet forming and drying

Form a sheet and couch it onto a post (see pp. 99–100). Use tweezers to position the flowers in pleasing arrangements on the surface. Press, then dry by stacking between newspapers (see pp. 101–103).

variations

You can use any other variety of pansy flowers in the same way with beautiful results. You will find that some colors of petals retain their color better when pressed than others. You may also wish to experiment with tinting the paper pulp to complement the colors of your flowers (see p. 13).

Pressed violas laid on the surface of the paper preserve both their shape and color →

Cornflower

Centaurea

Simple

Once considered a weed found growing among corn, cornflower now appears wild on unsprayed land. The juice from the flowers is sometimes mixed with alum to make a blue dye used by watercolorists. Cornflower is also known as bluebottle or blavers.

equipment

Basic papermaking equipment (see pp. 16—18).

cultivation

Cornflower is an annual that likes an open sunny situation and will grow in any well-drained soil. Spring-sown seeds flower in late summer. Fall-sown plants need some protection in winter for an early crop of flowers. The plantlets do not transplant well so make sure you choose the sowing position carefully. Flowering begins about eight weeks after seeds are sown.

harvesting and preparation

Use the flowers only. Pick the flower heads regularly to encourage new growth. Pull the flower heads apart to separate the florets and spread them on a sheet of absorbent paper. Press and dry (see p. 11) and build up a good stock.

beating method

No extra beating is required once the pulp is prepared (see p. 98).

sheet forming and drying

Form a plain sheet and couch it onto a post (see pp. 99—100). Use tweezers to place individual florets in a pattern of your choice. Press, then dry between newspapers (see pp. 101—103).

variations

For a different effect try lightly blending some of the dried petals with the prepared pulp to make a flecked background for the florets.

Cornflower florets scattered over the paper's surface give a summer-time feeling

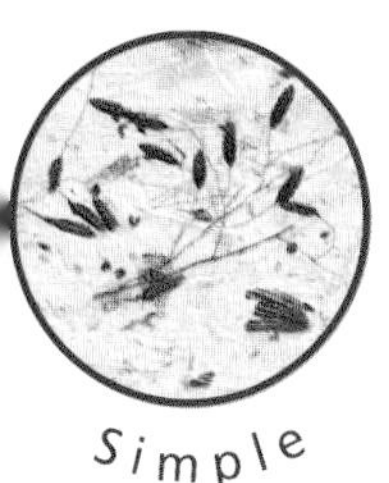

Simple

Wallflower

Cheiranthus

It is said that fragrant wallflowers were often planted high on the walls of castles and manor houses, so that the scent could waft through the windows of the bedchambers. They are sometimes called milkmaids.

equipment

Basic papermaking equipment (see pp. 16—18).

cultivation

The wallflower is usually biennial, but will flower as a perennial if left in the bed. It will grow in any well-drained soil. Sow seeds in late spring in open ground. When the seedlings are large enough to handle plant them out into a nursery bed, at 6-in (15-cm) intervals. In autumn set the plants out in their final position, at 12-in (30-cm) intervals.

harvesting and preparation

Use the flowers only. Although you can press and dry the flowers (see p. 11) you can also use fresh blooms, which will retain their color better.

beating method

Add a quantity of fresh or dried flowers to the prepared pulp and blend lightly enough to disperse them without breaking them up (see p. 98). Try not to make the concentration of flowers in the pulp too heavy; if the petals in the paper have space around them a pleasant "bleeding" aura will be visible.

sheet forming and drying

Form a sheet and couch it onto a post (see pp. 99—100). Press, then dry hanging up (see pp. 101—103). This pulp will stain the felts, so do not couch it in a post with any other papers.

A variation made with pressed flowers shows less bleeding

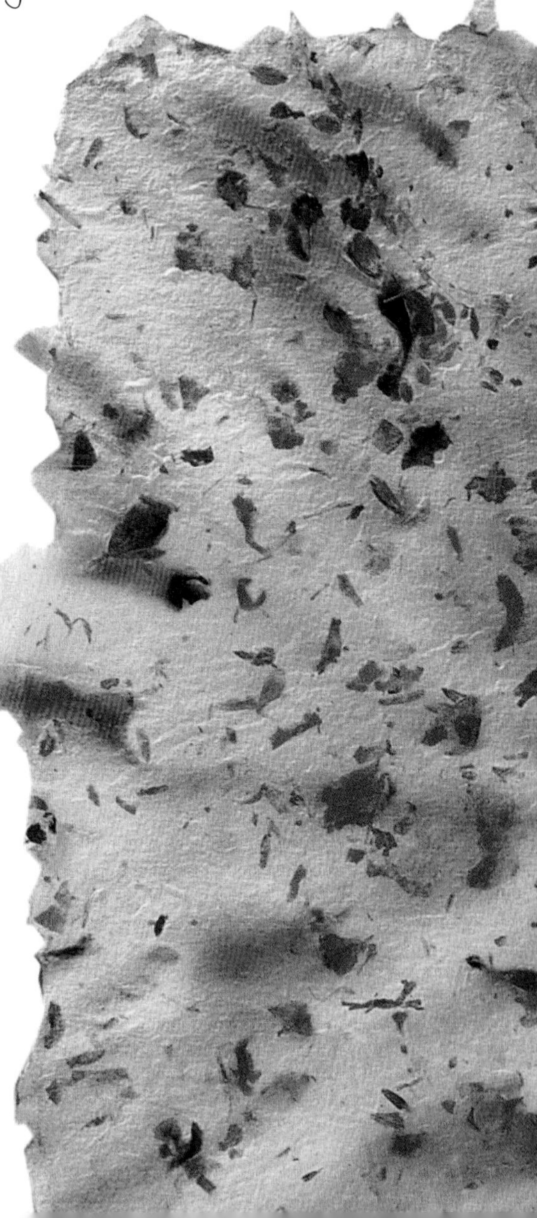

Fresh wallflowers bleed a lot of color into the paper pulp

Love-in-a-Mist

Simple

Nigella damascena

Love-in-a-mist is also known as devil-in-the-bush. The Latin name derives from "niger," referring to the blackness of the seeds, which are used in Mediterranean countries to decorate the crusts of loaves.

equipment

Basic papermaking equipment (see pp. 16—18).

cultivation

Love-in-a-mist is a self-seeding annual that will flourish in any garden soil and likes a sunny position. Sow in fall, or in spring for later flowering. Thin the seedlings to stand 8 in (20cm) apart. If you are a flower arranger, you may like to keep the seedpods, as they are spectacular.

harvesting and preparation

Use the flower heads, the fern-like leaves, and the seeds. Gather flowers throughout the season, leaving some on the plant to form the seed. Press and dry the flowers and leaves (see p. 11). The seeds are contained in a papery, balloon-like case. Take them from the case, use some for paper-making, and keep some for resowing.

beating method

Sprinkle some seeds into the prepared pulp and stir (see p. 98).

sheet forming and drying

Form a sheet and couch it onto a post (see pp. 99—100). Use tweezers to position the dried flowers and their "mist" of leaves on the surface. Press, then dry by stacking between newspapers (see pp. 101—103).

variations

Love-in-a-mist usually has blue flowers, but one variety, Persian Jewels, has a multicolored mix of flowers, which you can use in paper.

The delicate tendrils that frame the flower heads make a beautiful pattern in the paper

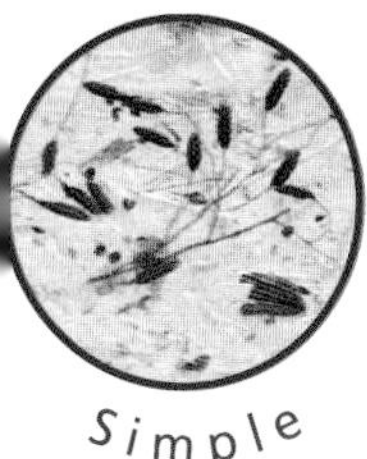

Tulip Petals

Tulipa

First cultivated in Turkey, where wild tulips grow, their name derives from the Arabic word for turban. Tulips have been immensely popular in Europe since they were first imported in the 16th century, and to this day Holland is the main center for cultivar production.

equipment

Basic papermaking equipment (see pp. 16–18).

cultivation

Tulips are bulbous plants which will grow in any ordinary, well-drained garden soil, either in full sun or light shade. Plant in late fall at a depth of about 4 in (10cm) for spring flowers.

harvesting and preparation

Use the petals only. When the flower begins to show color cut the stems and enjoy the tulips as cut flowers. When you are ready to make the paper, pick the petals and use them fresh, or dry them in a net bag (see p. 11) for future use.

beating method

Add fresh or dried petals to the pulp and blend lightly (see pp. 98). Leave the mixture in the vat until some color develops throughout the pulp.

sheet forming and drying

Form a sheet and couch it onto a post (see pp. 99–100). Press, then dry hanging up (see pp. 101–103). This pulp will stain the felts.

variations

You can use any variety of tulip: choose your favorite colors and petal shapes. Instead of growing your own tulips you can buy the flowers from a florist nearly all year round.

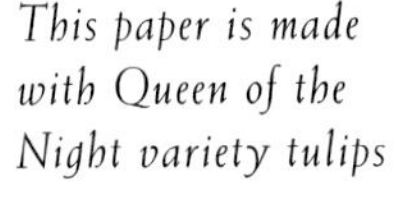

This paper is made with Queen of the Night variety tulips

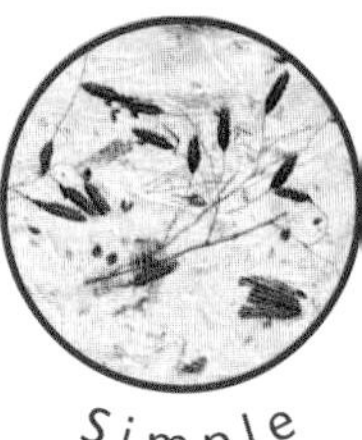

Simple

Borage & Daisy

Borago officinalis and *Bellis perennis*

Borage is also known as bee bread, coll tankard, or hero of gladness. Borage flowers and leaves are sometimes added to summer drinks to impart a cucumber flavor. They are often frozen in ice cubes for this purpose, but can be preserved in the same way for papermaking.

equipment

Basic papermaking equipment (see pp. 16—18).

cultivation

Borage is an annual that likes an open, sunny, and well-drained position. Sow seed in spring. Borage usually self-seeds. Daisy is a hardy perennial that flowers all summer and likes a very well-drained soil in sun or semi-shade. Propagate by division or sow seed in early summer.

harvesting and preparation

Collect the flower heads throughout the flowering season and press and dry them (see p. 11).

beating method

When preparing the basic pulp (see p. 98), add some colored card to give it a blue tint.

sheet forming and drying

Form a sheet and couch it onto a post (see pp. 99–100). Use tweezers to position the flowers on the surface. Press, then dry by stacking between newspapers (see pp. 101—103).

The cool effect of a blue background sets off the flowers, evoking the mood of a summer evening →

← *Common lawn daisies will press flat, but with ornamental daisy varieties (which are actually chrysanthemums) you could just use the petals*

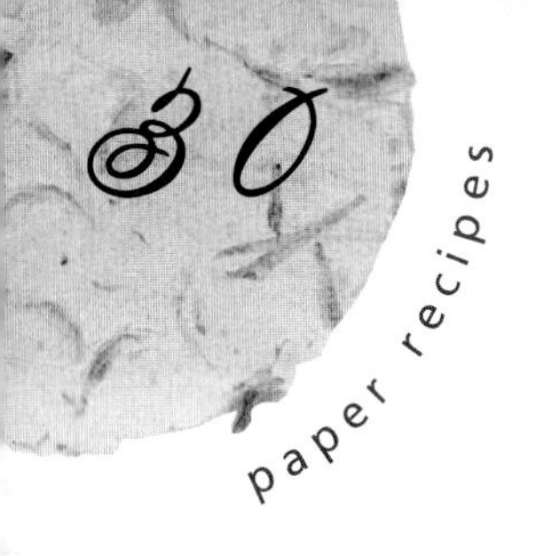

Marigold

Calendula officinalis

Marigold petals retain their color beautifully, and the dried petals on the surface of the sheet gleam enchantingly. The Latin name is derived from the belief that this plant was always in flower on the first day of each month. They are also known as Mary buds, gold of ruddes, or pot marigolds.

equipment

Basic papermaking equipment (see pp. 16–18).

cultivation

Marigold is a hardy annual and very easy to grow. It likes a sunny position in well-drained soil; too rich a soil leads to excessive leaf growth at the expense of flowers. Sow seeds in autumn, with some winter protection, to flower in spring and summer. Another sowing in summer will bring late flowers and an early spring bloom. Marigold will usually self-seed and transplants well.

← Using fewer petals in the pulp makes more of a feature of the pressed petals

harvesting and preparation

Use fresh and dried petals. Collect the flower heads and remove the petals. Press some of the best shaped ones between sheets of absorbent paper for at least 24 hours (see p. 11). Dry the remaining petals (see p. 11).

beating method

Stir the unpressed petals into the prepared pulp (see p. 98).

sheet forming and drying

Form a sheet and couch in onto a post (see pp. 99–100). Use tweezers to position the pressed petals in an arrangement of your choice on the ace. Press, then dry hanging up (see pp. 101–103).

This paper uses calendula petals but you could experiment with other types of marigold →

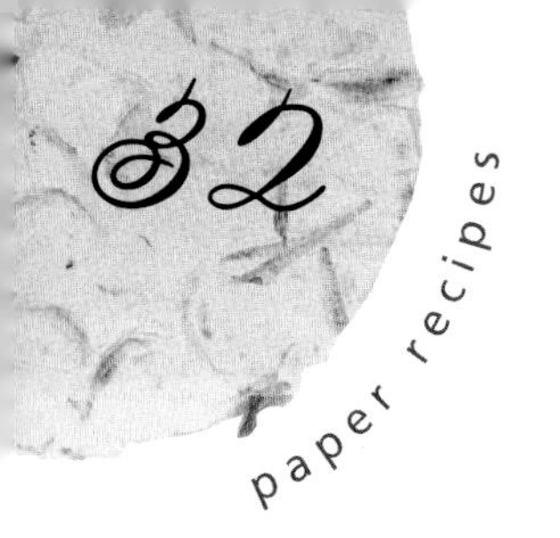

Nemophila

Nemophila maculata

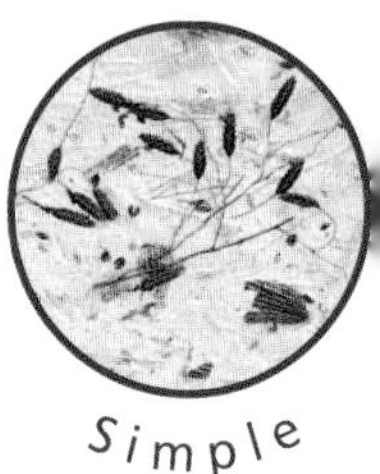

Simple

Nemophila, also known as baby blue eyes, is such a useful, pretty annual for the front of a border that it deserves to be grown widely. Another very similar plant is the California bluebell (Hydrophyllaceae).

equipment

Basic papermaking equipment (see pp. 16—18).

cultivation

Nemophila is a hardy annual that grows to about 6 in (15cm) in height. Grow in sun or semi-shade in fertile, well-drained soil. It will flower all summer if sown at intervals starting as soon as the soil is warm enough. Sow the seed where it is to flower and thin seedlings to 4 in (10cm) apart.

harvesting and preparation

Use the flowers only. Remove the flower heads in season. Press and dry them (see p. 11).

beating method

No extra beating is required once the pulp is prepared (see p. 98).

sheet forming and drying

Form a sheet and couch it onto a post (see pp. 99—100). Use tweezers to position the flower heads in pleasing arrangements on the surface. Press, then dry by stacking between newspapers (see pp. 101—103). This pulp will stain the felts.

variations

For a different effect try lightly blending some fresh flowers with the prepared pulp to make a flecked background for the surface flowers.

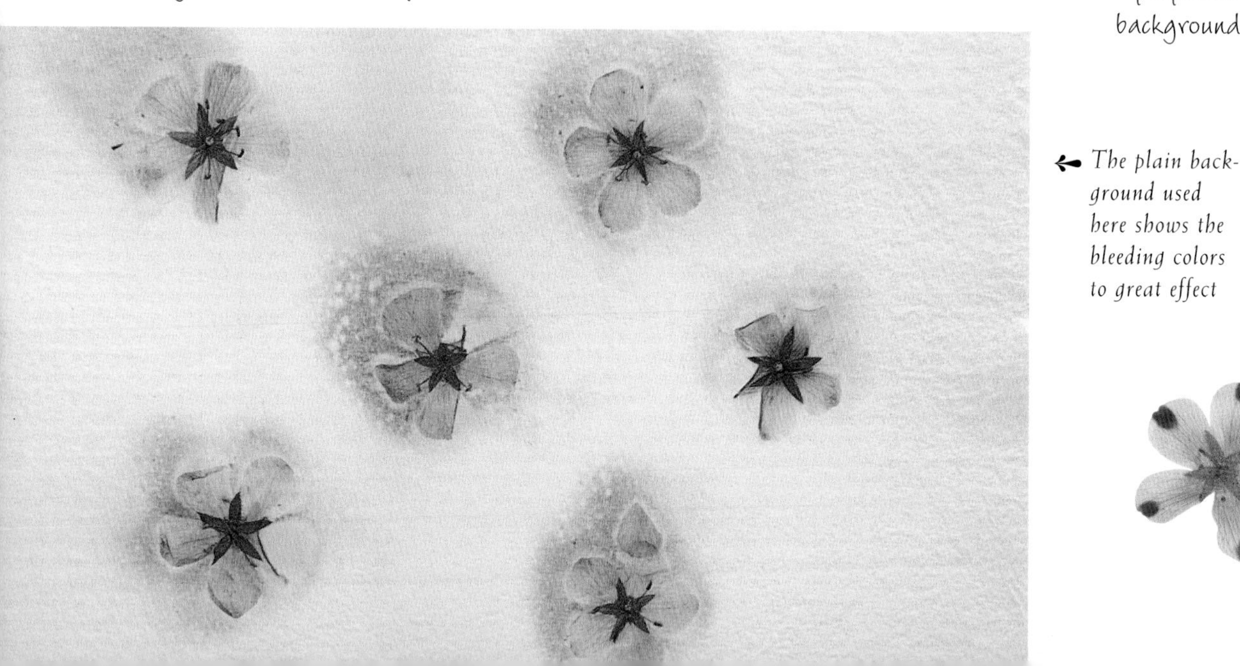

← *The plain background used here shows the bleeding colors to great effect*

Rose Petals

Rosa gallica officinalis

Roses, also known as gift of the angels, have appeared in art for centuries, in Minoan frescoes, the decoration of European cathedrals, and in early Italian paintings. They have many culinary and medical uses — rose essence is among the safest healing substances known.

equipment

Basic papermaking equipment (see pp. 16—18).

cultivation

Roses can be grown successfully on a wide range of soils, provided they are not subject to waterlogging in winter. They like an open, sunny site. In summer the dead flowers should be removed and the plant cut back to a wood bud.

harvesting and preparation

Use about 1 ounce (25g) of either fresh or pressed and dried petals (see p. 11). When harvesting, sort the petals into groups which have different effects; for example dark reds are the most dramatic, whereas white petals tend to turn brown-beige in paper and are better set aside for drying or using in pot pourri. Tiny buds can also be pressed and used in the petal paper.

beating method

Add some petals to the prepared pulp (see p. 98) and blend to disperse fragments of petal through the mixture. Add the remaining whole petals to the vat and stir.

sheet forming and drying

Form a sheet and couch it onto a post (see pp. 99—100). Press, then dry hanging up (see pp. 101—103). This pulp will stain the felts.

variations

You can use any color of rose petals that you have available. Red petals have a tendency to "bleed," producing an attractive "blush" in the sheet.

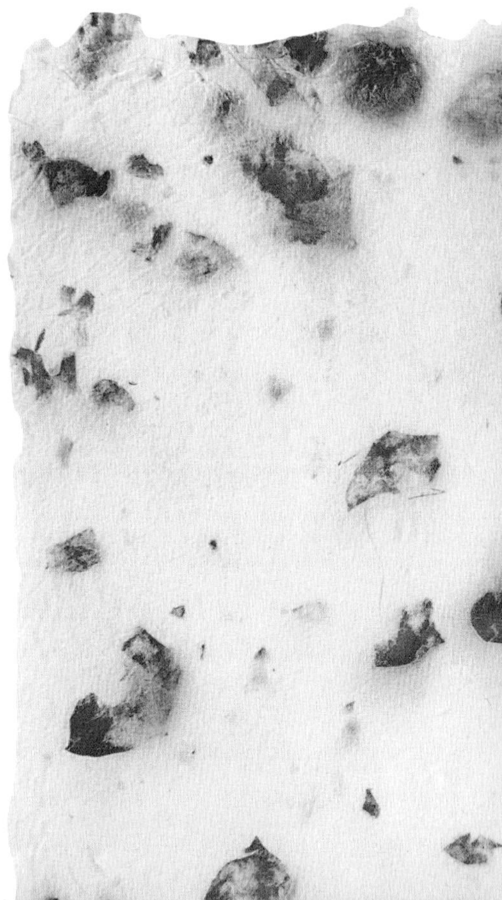

Fragments of petals give a faint pink blush, which spreads through the paper

Larkspur & Hydrangea

Simple

Delphinium and Hydrangea macrophylla

The proper name for larkspur, delphinium, comes from "delphis," the Greek for dolphin, alluding to the shape of the flowers. Hydrangea also derives its name from Greek — "hydra" means water and "aggros" means a jar — describing the shape of the seed capsules.

equipment

Basic papermaking equipment (see pp. 16–18).

cultivation

Larkspur grows in a moist, deep soil in an open situation. It needs to be staked and sheltered from strong winds. It does not transplant well so choose a position carefully before sowing seeds in either fall or spring. Larkspur will often flower twice a year.

Hydrangeas are deciduous bushy shrubs that like fertile, well-drained soil. Blue-flowered varieties need an acid soil or bluing powder can be added every ten days prior to flowering. The color can be more easily controlled when hydrangeas are grown in tubs. Propagate by softwood cuttings in summer.

harvesting and preparation

Use the flowers only. Cut the whole stalk of the larkspur and hang in bunches for three or four weeks to dry. When dry store the flowers in a box until needed.

Remove blue hydrangea flower heads at the peak of their development. Separate the individual florets and press and dry them (see p. 11).

beating method

Lightly blend the larkspur flowers into the prepared pulp (see p. 98).

sheet forming and drying

Form a sheet and couch it onto a post (see pp. 99–100). Use tweezers to position the hydrangea florets in a pleasing arrangement on the surface. Press, then dry by stacking between newspapers (see pp. 101–103). This pulp will stain the felts.

This variation uses blue larkspur florets in the pulp to make a background for the blue hydrangea flower heads

This paper uses blue larkspur in the pulp and both flowers as pressed features

Primrose

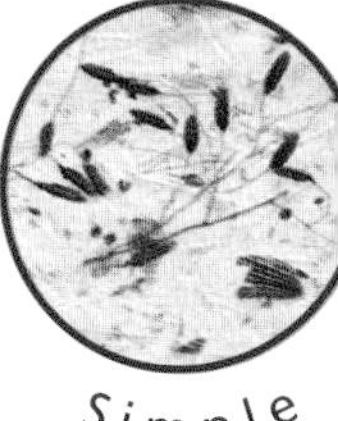
Simple

Primula vulgaris

The fragments of stalks, leaves, and flowers in the background give a lively, spring-like feeling to the paper, while the whole flower heads remain recognizable. A creamy base pulp is especially suitable for this paper. The primrose is also called a "spinkie" in some places.

equipment

Basic papermaking equipment (see pp. 16—18).

cultivation

Primroses thrive in cool, good, loamy, leafy soils and will survive in semi-shade. Sow seed in spring. Increase plants by division after flowering. Primroses last for up to 25 years and picking them does no harm, provided the roots are left undisturbed.

harvesting and preparation

Use fresh flowers, leaves, and stalks for the background and set aside some well-shaped flower heads for use on the surface of the paper.

beating method

Add the fresh flowers, leaves, and stalks to the prepared pulp (see p. 98) and blend in short bursts until you achieve the desired background density.

sheet forming and drying

Form a sheet and couch it onto a pos (see pp. 99—100). Use tweezers to position the pressed and dried flowers Press, then dry by stacking between newspapers (see pp. 101—103).

variations

Any of the primula family can be use in papermaking and they all have lovel flowers for pressing. There are also bo varieties and pot plants that are available all year round.

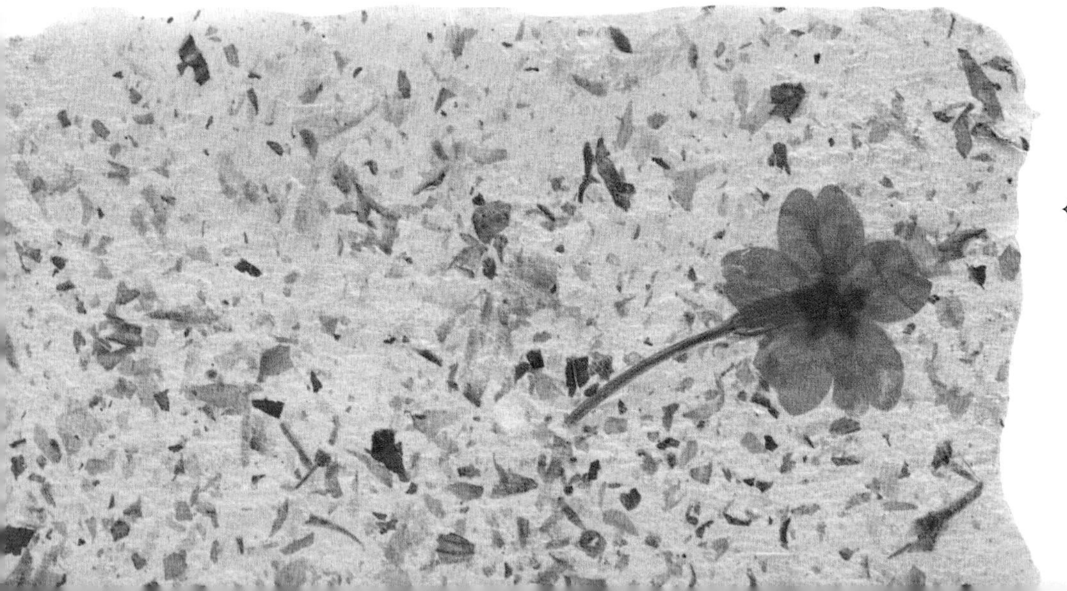

One option is to use plenty of leaf material in the pulp and make a feature of a single pressed primrose

This paper uses leaf and flower fragments quite sparingly in the pulp, but uses more pressed flowers for extra color

Pot Pourri

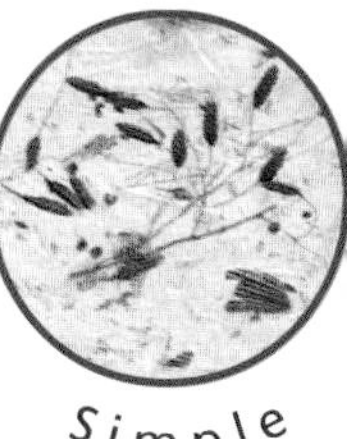

Simple

This is a beautiful paper that can be made at any time of the year and is a happy reminder of a summer garden. You can make your own pot pourri from the plants growing in your garden, allowing you to choose your favorite scents, colors, and shapes to add into the mix.

equipment

Basic papermaking equipment (see pp. 16—18) and coffee grinder.

cultivation

Any combination of scented leaves, flowers, and buds from your garden can be used. Traditionally, lavender and rose petals are the main ingredients.

harvesting and preparation

Use flowers, leaves, and buds. Collect and dry (see p. 11) each flower in its season and make up the mixture when all the ingredients are dry. Seeds and spices can be ground up and added. Fixative powders such as orris root (Iris germanica), violet root (Viola odorata), and sweet flag (Acorus calamus) can be added to make the pot pourri long lasting. You may choose to use bought pot pourri, some of which contain hard items that need grinding in a coffee grinder before being used for paper. About 4 ounces (100g) will be plenty for 12 sheets.

beating method

Add the dried mixture to the prepared pulp (see p. 98) and blend it a little to hydrate it and bring the pulp to an even consistency.

sheet forming and drying

Form a sheet and couch in onto a post (see pp. 99—100). Press, then dry hanging up (see pp. 101—103). This pulp will stain the felts.

Variations

Although the perfume is strong when you are making the paper, unfortunately the scent does not last long in the dried paper. You may want to use extra scent (see p. 109).

← Homemade pot pourri with lots of red rose petals produces a very pink paper

Store-bought pot pourri gives good overall color →

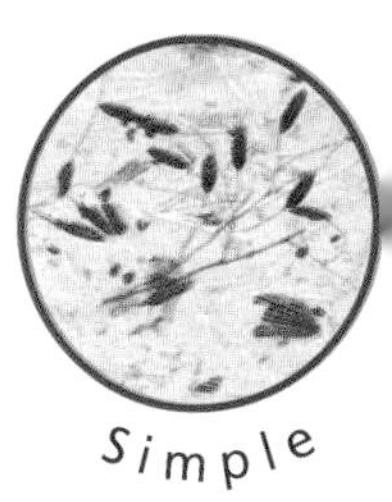

Simple

Elderflower

Sambucus

The botanist Richard Mabey summed up the mixed merits of the elder as follows. "It is too small to be a tree yet too large and airy for a bush. Its roots and heartwood are as hard as ebony yet the young branches are soft and weak. The umbels of its white flowers smell of honey, the leaves of mice nests . . ."

equipment

Basic papermaking equipment (see pp. 16—18).

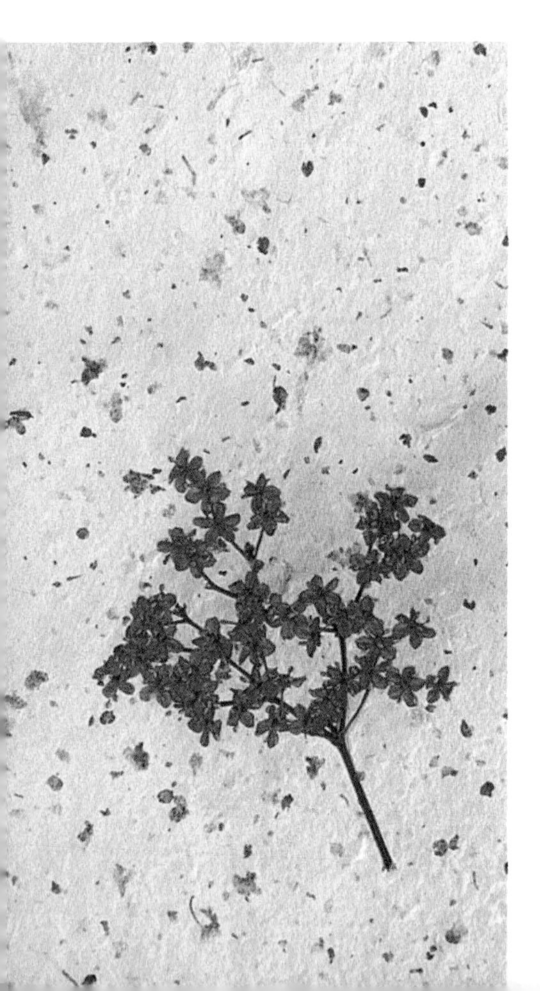

cultivation

As a perennial deciduous shrub, many varieties grow wild extensively in Britain and Europe, but cultivated forms can grow in gardens elsewhere. Elderflower is also used as an ingredient in a muscatel-flavored country wine.

harvesting and preparation

Use the flowers only. Pick the flowers when they are fully mature and shake the florets off the stalks.

beating method

Blend the florets lightly into the prepared pulp (see p. 98), mixing them in rather than breaking them up.

The tiny flowers of an elderflower head form a delicate paper

sheet forming and drying

Form a sheet and couch it onto a post (see pp. 99—100). Position a sprig of elderflower on the surface with tweezers. Press, then dry between newspapers (see pp. 101—103).

variations

Small flowers from other shrubs such as some Viburnums and some Spireas may also be used. Like most white flowers, they oxidize quickly and turn brown in the finished paper.

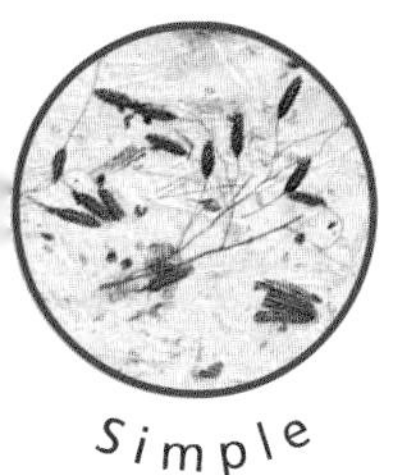

Simple

Fennel Leaves

Foeniculum vulgare

Fennel, sometimes called sweet fennel, was one of the nine herbs held sacred by the Anglo-Saxons, who believed it had protective powers against evil. It is a tall, feathery plant equally valued for its graceful form and its usefulness in cooking. You can also use dill leaves for a similar looking paper.

equipment

Basic papermaking equipment (see pp. 16–18).

cultivation

Fennel is a hardy biennial or perennial that thrives in a sunny position in fairly rich, well-drained soil. Sow in late spring to early summer. Fennel self-seeds when established but to propagate divide in fall. Do not grow fennel near dill or coriander as the plants will cross-pollinate.

harvesting and preparation

Use fresh leaves. Gather and roughly cut the leaves as you need them.

beating method

Blend the cut fresh leaves into the prepared pulp (see p. 98). You do not want to lose the fern-like effect of the leaves, so avoid overbeating.

sheet forming and drying

Form a sheet and couch it onto a post (see pp. 99–100). Press, then dry by hanging up (see p. 101–103).

The ferny leaves give a subtle effect in the paper

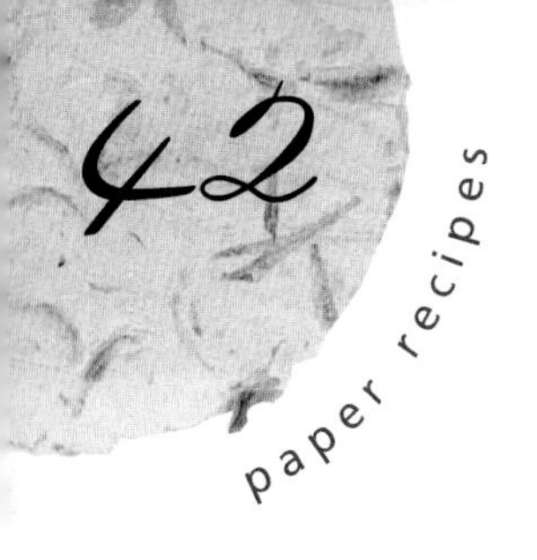

Simple

Parsley

Petroselinum crispum (umb.)

Legend has it that the parsley seed goes to the Devil and back nine times before germinating. The delicate green speckle of parsley looks particularly good blended with a cream-colored pulp. There are different types of parsley, and this variety is known as moss curled parsley or French curly leaved parsley.

equipment

Basic papermaking equipment (see pp. 16—18).

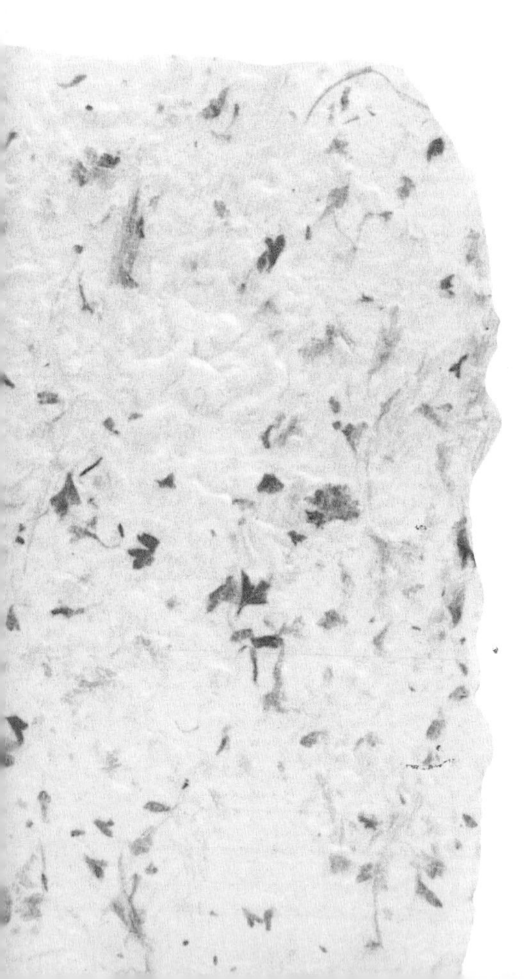

← *Air drying gives a textured effect*

cultivation

Parsley is usually grown as an annual, although strictly speaking it is biennial. It likes a rich fertile soil with good tilth and makes a good edging plant for a herb or vegetable garden. The seeds are very slow to germinate (6—8 weeks). To encourage them, soak overnight and sow into a small furrow wetted with a trickle from a boiling kettle. Parsley can be container grown and kept indoors for the winter to produce fresh leaves. It does not, however, dry well.

harvesting and preparation

Use fresh leaves. Cut whole leaves from the plant just prior to use.

beating method

Add a bunch of fresh leaves to the prepared pulp (see p. 98). You do not want to break up the leaves too much so blend with a short burst of just 2—5 seconds, enough to leave some frilly bits of leaf visible. Repeat, adding leaves until you reach the desired intensity.

sheet forming and drying

Form a sheet and couch it onto a post (see pp. 99—100). Press, then dry by hanging up or by air drying (see pp. 101—103).

Fresh leaves are vividly green where they show at the surface of the paper →

Scented Geranium

Simple

Pelargonium

A few rose- or lemon-scented geranium leaves placed in the bottom of a cake tin before the sponge mix i poured in will add flavor and make an imprint on the cake, just as they will make an imprint on your paper if you peel the leaf off after pressing.

equipment

Basic papermaking equipment (see pp. 16—18).

cultivation

Geraniums are frost-tender perennials that like a sunny position in well-drained neutral to alkaline soil. They can be easily propagated from softwood cuttings from spring to fall. They are usually grown in pots, either as house-plants or so they can be brought indoors in winter. Do not over-water.

harvesting and preparation

Use both fresh and dried leaves. For maximum perfume pick the leaves just before the flowers open. Divide them into two groups and press and dry one group (see p. 11).

beating method

Blend the fresh leaves into the prepared pulp (see p. 98) for as long as it takes to reach the desired density of material.

sheet forming and drying

Form a sheet and couch it onto a pos (see pp. 99—100). Use tweezers to position the dried leaves on the surfac Press, then dry by stacking between newspapers (see pp. 101—103).

variations

There are many varieties of geranium that have distinctive scented leaves. You can choose lemon, apple, rose, pine or rose-peppermint. Unfortunately th scent will not last for long and you ma want to use some other scenting techniques (see p. 109).

Peeling a leaf off the surface after it has been dried leaves an impressed image

You can choose geranium leaves for their scent, shape, or color

Maple Leaf

Acer

Simple

Maples come in many forms and colors, from the familiar stylized emblem of the Canadian flag to the finely divided leaves of many Japanese varieties. You can use the leaves of any maple when they color in the fall, or you could grow the red-leaved Acer palmatum 'Dissectum Atropurpureum'.

equipment

Basic papermaking equipment (see pp. 16—18).

cultivation

The acer family of trees and shrubs grow in any fertile, moist soil in sun or partial shade.

harvesting and preparation

Use dried leaves and bark. Pick the leaves at any stage in their development and press and dry them (see p. 11). When you are ready to use them, float the leaves in a bowl of water with a drop of detergent for 1 hour, to help them lie flat and adhere to the paper sheet. The bark naturally comes loose and can be gathered and dried in a net bag (see p. 11).

beating method

Lightly blend the bark into the prepared pulp (see p. 98). The color of the bark and leaves slightly dyes the whole sheet with a warm tinge.

For a subtle effect, emboss the paper with the imprint of a leaf

sheet forming and drying

Form a sheet, couch it, and while still wet use tweezers to position the dried leaves. The leaves will emboss the sheets so they cannot be pressed in a post. Press each sheet between boards, then dry between newspapers (see pp. 101—103). Thick leaves sometimes lift off the sheet, so a touch of craft glue may be used to secure them.

variations

The embossed sheet shown here is made by placing an unsoaked leaf on the wet couched sheet. Press and dry, then lift it off, leaving only its impression.

The bark gives an all-over color, while the leaves make a strong pattern

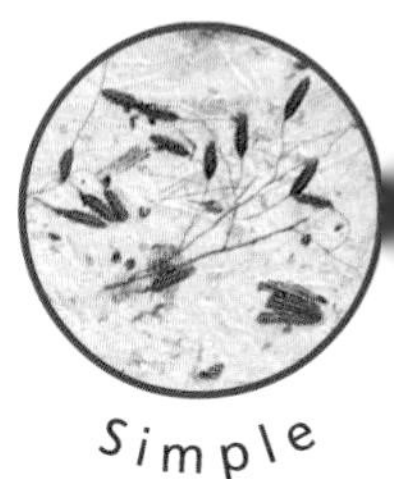

Ivy

Hedera

The ivy is dedicated to Bacchus, the ancient Greek god of wine. Though variable in shape, the leaves always have five lobes and are placed alternately along the stems, and are evergreen.

equipment

Basic papermaking equipment (see pp. 16—18) and a pair of pruning shears .

cultivation

Ivy grows wild extensively, both as a climber on trees and on its own. There are many cultivated forms producing variegated foliage. They need a poor, alkaline soil and more light than the wild forms, to bring out the leaf color. Propagate in late summer by cuttings or rooted layers.

harvesting and preparation

Gather the leaves and tendrils at any time and press (see p. 11).

beating method

A green pulp is made by adding scraps of green paper to the mix. Beat as usual (see p. 98).

sheet forming and drying

Form a sheet, couch it and while still wet position the ivy tendrils with tweezers. The leaves will emboss the sheets, so they cannot be pressed in a post. Press each sheet between boards, then dry by stacking between newspapers (see pp. 101—103).

Whole stalks or single leaves can be arranged in patterns of your choosing

Simple

Corn Cob Silk

Zea mays

The forbears of maize or sweet corn were growing in Mexico earlier than 5,000 BC. Columbus took Indian maize to Spain from Cuba, and from there it spread through the Mediterranean and beyond.

equipment
Basic papermaking equipment (see pp. 16—18).

cultivation
Sow seeds of this annual in late winter or early spring, in peat pots at 12°C (55°F). Transplant in late spring to an open site where they will receive the greatest possible amount of sun. Grow in blocks, so the pollen from the flowers drops on to the silks around the cobs.

harvesting and preparation
Use the silks saved from the corn cobs. Do not remove the silk before the cob is mature. Keep the rest of the cob to cook and spread the silks out in an airy position to dry (see p. 11). Store them in a paper bag.

beating method
Cut the silks into tiny pieces and blend lightly into the prepared pulp (see p. 98).

sheet forming and drying
Form a sheet and couch it onto a post (see pp. 99—100). Press, then dry by hanging up (see pp. 101—103).

The fine texture of this paper makes it good for writing on, but you will need to add size (see p. 98)

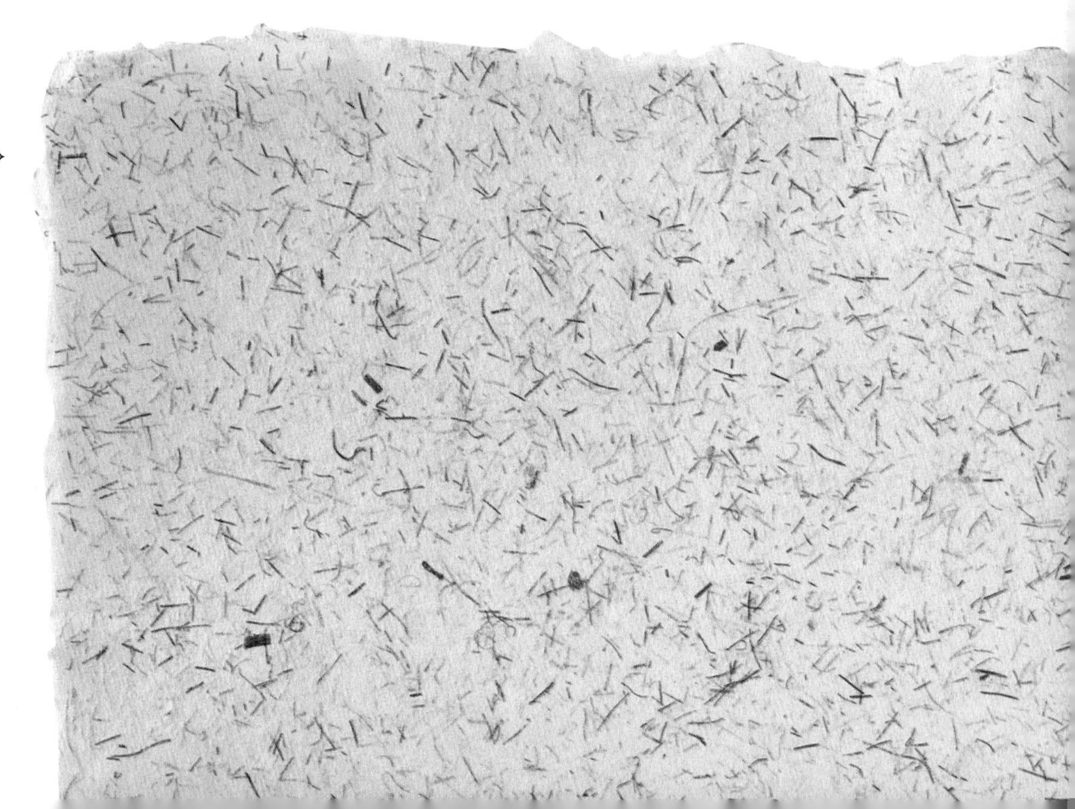

Skeleton Leaves

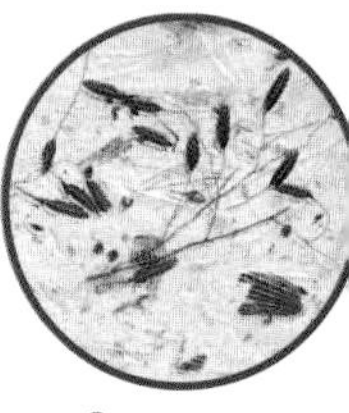

Simple

Some fallen tree leaves decay gradually on the woodland floor and reach a point when the soft part of the leaf disappears, leaving the skeleton of veins. The best time to search for them is in late fall or winter. You can also buy perfect skeleton leaves from craft stores.

equipment

Basic papermaking equipment (see pp. 16—18).

harvesting and preparation

Magnolia, holly, sycamore, and some poplar leaves are particularly good. Gather leaves which have fallen and are starting to decompose, leaving just the veins. When you are ready to use them, place the leaves on the surface of a bowl of warm water for about 1 hour, with a drop of detergent to aid wetting.

beating method

Prepare a batch of plain pulp (see p. 98). No extra beating is required.

sheet forming and drying

Form a plain sheet, couch it (see pp. 99—100) and while still wet position leaves on the surface with tweezers. The leaves will emboss the sheets, so cannot be pressed in a post. Press each sheet between boards, then dry between newspapers (see pp. 101—103).

variation

If you are using very pronounced leaves you may want to couch a plain, thin sheet on top of the decorated sheet before pressing and drying (see laminating technique, p. 107).

A variety of leaf shapes are found in my yard each winter

These perfect leaves were supplied by a craft store

Simple

Using different grass varieties will give you an amazing array of papers, but remember that grasses with very fat seedheads will not be suitable.

equipment

Basic papermaking equipment (see pp. 16—18).

cultivation

There is an enormous range of ornamental grasses cultivated as hardy annuals. To grow a particular variety sow some seeds in spring in well-drained soil in a sunny, open position where they are to flower. They may need thinning into clumps.

harvesting and preparation

You can use both seed heads and leaves. As soon as the grasses have flowered, cut the seedheads, keeping long stems. Use fresh or store in bunches. When needed, cut them into 1/2 in (13mm) lengths.

beating method

Blend according to the size of the seed — larger seeds will need longer blending to gain an even distribution. Add the cut leaves and blend together until the prominent pieces are reduced.

sheet forming and drying

Form a sheet and couch it onto a post (see pp. 99—100). Press, then dry by hanging up (see pp. 101—103).

The relief texture of this paper makes it a good candidate for covering notebooks or albums, or for collages

Simple

Bulrush

Typha latifolia

Bulrush is also known as brown busbies, reed mace, or cattail. This delicate paper has an all-over effect, which is ideal for endpapers. A "sized" version can be written upon (see p. 98).

equipment

Basic papermaking equipment (see pp. 16–18).

cultivation

The perennial bulrush commonly grows wild in ditches and at the edges of ponds and lakes. It can be cultivated and grown in water or wet ground in sun or partial shade. The larger species should be grown in containers as they tend to "overwhelm" other plants. Plant in fall or spring. Propagate by division in spring.

harvesting and preparation

Use the seeds only. Gather four rusty-brown flower spikes before they burst. Invert each spike in a large paper bag and gradually break it apart, letting the paper bag catch the thousands of seeds housed within the spike. Store in a dry place until needed.

beating method

Add the seeds to the prepared pulp and blend lightly (see p. 98) to distribute them evenly.

sheet forming and drying

Form a sheet and couch it onto a post (see pp. 99–100). Press, then dry by hanging up (see pp. 101–103).

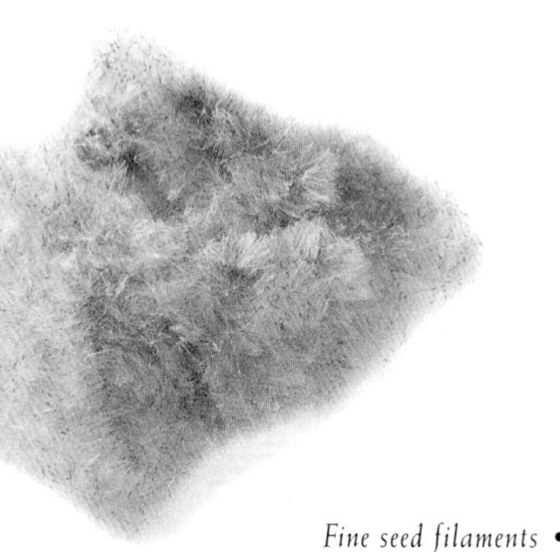

Fine seed filaments give a very smooth textured paper

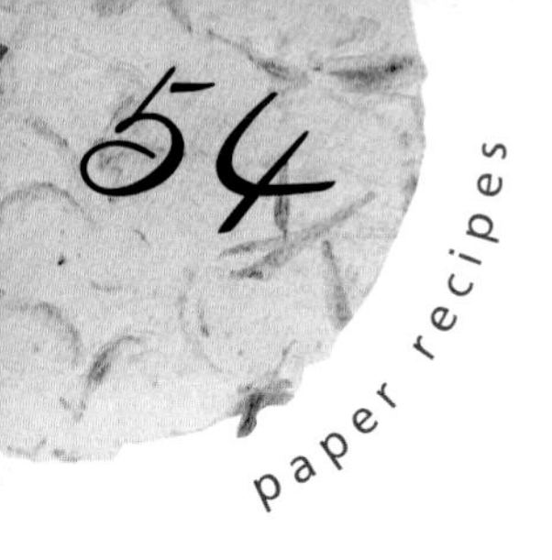

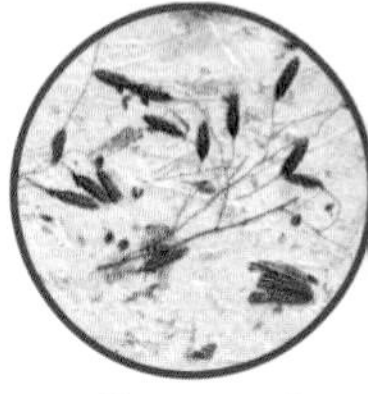

Simple

Salsify

Tragopogon porrifolius

To enjoy this flower before its petals fold, it is necessary to get up early on a sunny morning. The plant fragments make a bold pattern with patches of yellow that "bleed" out beautifully. Salsify is also known as goats' beard or vegetable oyster.

equipment

Basic papermaking equipment (see pp. 16—18).

cultivation

Salsify is a hardy biennial with an edible root and a beautiful purple flower which closes by noon. To grow salsify as a root vegetable you need a deep, rich soil, but to get just a lovely seed head — the part used in papermaking — salsify will grow on a stony site and even self-seed between paving stones. It prefers a sunny site.

harvesting and preparation

Use the seed heads. When the seed head has developed pick the plant by the stalk and leave in a vase without water until needed.

beating method

Stir the seed heads into the prepared pulp (see p. 98).

sheet forming and drying

Either form a sheet and couch it onto a post (see pp. 99–100) or air-dry it on the mold (see p. 103). Note: this pulp will stain the felts if you couch it.

The seeds within the pulp produce a high-relief texture

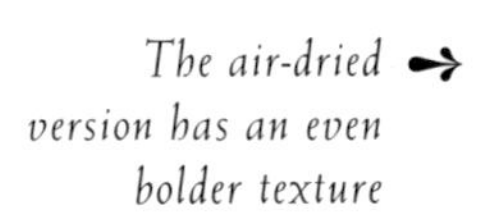

The air-dried version has an even bolder texture

Cow Parsnip

Simple

Heracleum

The robustness of this common weed — often just called hogweed — when growing in woodlands, grasslands, and at the roadside, led to it being dedicated in name to Hercules. The striking patterns created by the individual seeds give this paper great movement and life.

equipment

Basic papermaking equipment (see pp. 16–18).

cultivation

The biennial cow parsnip grows wild in Europe, northern Asia, and western North America.

harvesting and preparation

When the seeds are ripe enough they can be shaken from the plant and collected in a cloth positioned under the flower head. You can cut the stems if you wish to transport the whole heads but some people find the sap is a skin irritant. Dry the seeds (see p. 11).

beating method

Either add the seeds to the prepared pulp (see p. 98) and leave to soak for 24 hours before stirring to distribute them, or leave the seeds to stand for 1 hour in a bowl of hot water before stirring them into the pulp.

sheet forming and drying

Form a sheet and air-dry it on the mold (see p. 103).

This dense paper has seeds both in the pulp and also added to the surface →

← *This sheet uses fewer seeds, stirred into the pulp*

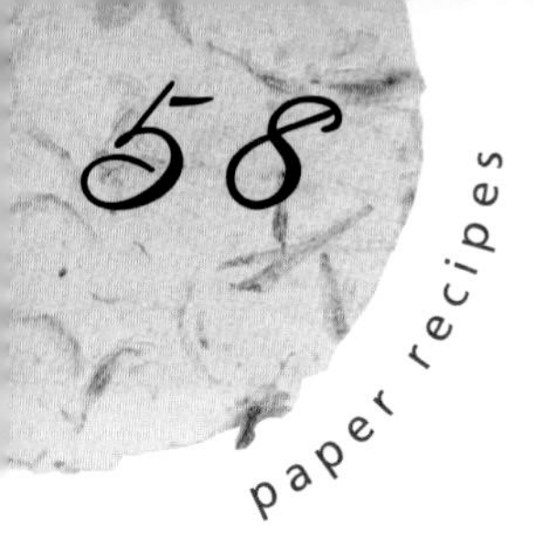

Linseed

Linum usitatissimum

The linseed plant has been of exceptional importance to mankind over the last 7,000 years. Linseed may be roasted and eaten, the oil is a constituent of varnishes and linoleum, and the flax fibers extracted from the stems are used to make linen.

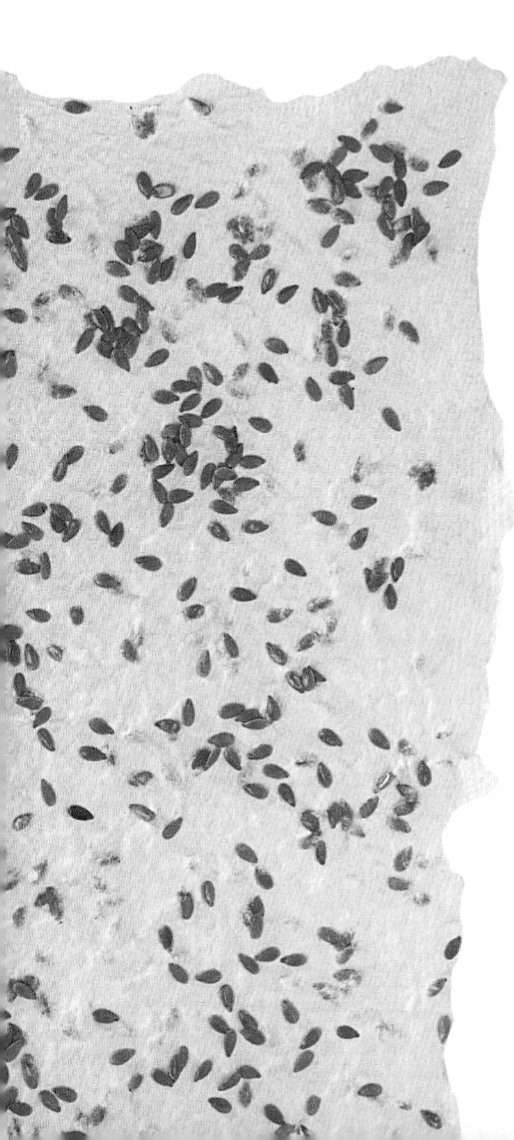

equipment

Basic papermaking equipment (see pp. 16—18).

cultivation

The flax plant has pretty blue flowers and can be grown in a garden border as an annual or perennial, but yields small seeds which are not ideal for papermaking. You can buy seeds at your nursery or garden store. I prefer to buy large seeds from a food reform or health store, where they are sold as a laxative or for making poultices to treat boils.

harvesting and preparation

None required.

← *The warm color and strong texture of this paper make it a great addition to any collage*

beating method

Stir most of the seeds into the prepared pulp (see p. 98), distributing them well.

sheet forming and drying

Form a sheet (see p. 99—100) and, while it is still wet, sprinkle the remaining seeds over the surface before air-drying on the mold (see p. 103). Note that if you attempt to couch and press this paper, there is a risk of the seeds sticking to the felts.

variations

The stems of the linseed plant, picked before the plant goes to seed, are used to produce linen thread. You can also make paper with these stems using the willow bast recipe (see p. 92).

Wych Elm Keys

Ulmus glabra

The Anglo-Saxon word "wych" meant pliable and refers to the useful quality of the tree's twigs. In this paper the keys seem to float decoratively across the olive-green flecked surface of the sheet.

equipment

Basic papermaking equipment (see pp. 16—18).

cultivation

Wych elm grows wild but, like many of the elm varieties, it is currently under attack by the virulent Dutch Elm disease fungus, so the elm keys may be difficult to find. There is hope that the health of these trees may soon be restored. Ulmus Americanus has similar keys.

harvesting and preparation

The fruit of the wych elm is an oblong seed pod (samara) with a single seed in the center. The pods are light and buoyant and float through the air. Collect them when you see them and use them fresh or dried (see p. 11). Keep the best specimens to one side.

beating method

Blend some of the keys into the prepared pulp (see p. 98), dispersing fragments into the mixture. Stir in the best specimens whole.

sheet forming and drying

Form a sheet and couch it onto a post (see pp. 99—100). Press, then dry by hanging up (see pp. 101—103). This pulp will stain the felts.

To create a richly patterned background, don't overblend the pulp

Parsnip Seeds

Simple

Pastinaca sativa

Parsnips have been cultivated since Roman times, and are still a popular winter vegetable. They are the one root vegetable that will stay quite happily in the ground all winter, not being damaged by frost. Any attractive seed may be used in the same way as this recipe to create an interesting paper.

equipment

Basic papermaking equipment (see pp. 16—18).

cultivation

Parsnips are a slow-growing root vegetable, so sow the seed as early as the weather allows in a deep, rich soil.

harvesting and preparation

Use the seed only. You can use seed from a packet bought from a nursery or garden store, as I did for this paper. Because parsnip seed needs to be fresh each year to germinate, you may have left-over seeds in the packet, which you can use for paper rather than wasting.

beating method

Lightly blend some of the seeds into the prepared pulp (see p. 98). Let it stand until it develops a slight coloration.

sheet forming and drying

Form a sheet (see p. 99) and sprinkle the remaining seeds over the surface. Air-dry it on the mold (see p. 103).

variations

If you grow your own parsnip and let it go to seed, the plant will develop a ferny top. You can gather the foliage, cut it into small pieces, and blend it into the pulp before adding the seeds.

← *Drying the paper on the mold gives it a very rough texture*

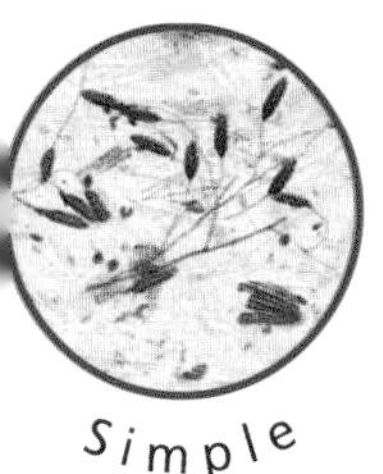

Simple

Seed Heads

Umbelliferae genus

Possible plants to use for this paper include parsley, caraway, cow parsley, chervil, and anise. Look around your neighborhood and keep an eye out for interesting shapes of seed head in late summer.

equipment

Basic papermaking equipment (see pp. 16—18).

cultivation

Use seed heads found in the wild.

harvesting and preparation

This paper can be made using the seed heads from many and various plants. A walk in the countryside in high summer will furnish you with plenty of material. Use the entire seed heads. Cut the seed heads from the plant before the seeds have become large and hard. Press and dry them (see p. 11).

beating method

No extra beating is required once the pulp is prepared (see pp. 98—99).

sheet forming and drying

Form a sheet and couch it onto a post (see pp. 99—100). Position the heads on the surface with tweezers. Press, then dry by stacking between newspapers (see pp. 101—103). This paper will stain the felts.

Take care when you press the seed heads, and you will be rewarded with delicate shapes →

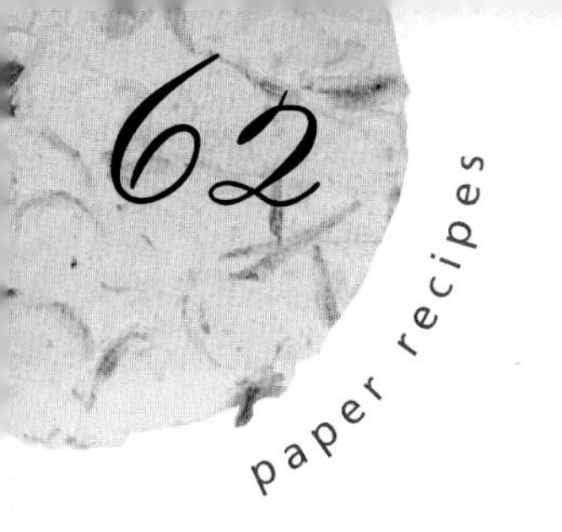

Simple

Honesty

Lunaria annua

In the language of flowers this plant symbolizes honesty and fascination. It is also known as dollar plant, satinflower, bread and cheese, and moon pennies — all names relating to the beautiful shiny oval seedpods which are so distinctive of this plant.

equipment

Basic papermaking equipment (see pp. 16–18).

cultivation

Honesty is a biennial which will grow in sun or shade and likes a well-drained soil. Sow seed in spring and, once established, honesty will set seed the next year and self-seed thereafter.

harvesting and preparation

The seed pods can be removed at different stages of development. At first they are green, they then develop a purple tinge, and turn white if fully mature. Remove the outside layers by rubbing the pod between finger and thumb, leaving the characteristic transparent window. The brown seeds may loosen and fall away in the papermaking process.

beating method

No extra beating is required once the pulp is prepared (see p. 98).

sheet forming and drying

Form a plain sheet, couch it, and, while still wet, position the honesty seeds with tweezers. Couch another plain sheet on top of the decorated sheet to laminate (see p. 107). The seeds will emboss the sheets, so cannot be pressed in a post. Press each sheet between boards, then dry between newspapers (see pp. 101–103). This pulp will stain the felts.

variations

For an interesting variation, leave the outside layers of the pod intact. This will produce some "bleeding" color in the sheet. Or emboss a sheet by pressing a pod into the wet, couched sheet, and removing it when the paper is dry.

← The inner and outer layers of honesty seeds are different colors

This laminated sheet → has both the imprint and color of honesty

Cooked papers

The recipes in this section involve cooking to one degree or another. For some recipes this may only involve making a tea, which is added to the standard batch of paper pulp (see p. 98). For the more adventurous papermaker, there are also recipes for making pure plant papers from fibers, leaves, vegetables, and fruit. These involve cooking with chemicals and therefore require additional equipment (see p. 19), space, and care (see pp. 104–105).

Safflower

Carthamus tinctorious

Safflower, also known as bastard saffron or dyer's thistle, has a stiff, whitish stem that branches at the top and carries orange flowers. The orange flower fragments complement the yellow stain they create in the paper pulp. Cloth dyers traditionally used safflower to make the red dye for cotton tapes that secured legal documents, and led to the use of the expression "red tape."

equipment

Basic papermaking equipment (see pp. 16—18).

cultivation

Safflower is a hardy annual that likes an ordinary well-drained soil and a sunny position. Sow seed in late spring.

harvesting and preparation

Use the flower heads. Pick the flower heads and dry them (see p. 11).

cooking

Pour 1 pint (600ml) of boiling water over 2 tsps of dried flower heads. Leave overnight for the color to develop.

beating method

Stir the plant material into the prepared pulp (see p. 98).

sheet forming and drying

Form a sheet and couch in onto a post (see pp. 99—100). Press, then dry hanging up (see pp. 101—103). This pulp will stain the felts.

In this sheet I used dried safflowers, bought from a dye store

Clematis & Leek

Clematis and Allium ampel

Clematis is also known as old man's beard or traveler's joy. In this paper the feathery white clematis seeds are set against a pure plant pulp made from the outer leaves of leeks.

equipment

Basic and adventurous papermaking equipment (see pp. 16—19).

cultivation

There are cultivated and wild varieties of the perennial clematis that have feathery, mop-like seed heads. All species like their roots in the shade in a well-drained soil. They climb to the sun, flowering at different times throughout the summer.

The leek is a member of the onion family, grown from seed in spring and transplanted in mid-summer, spaced 8 in (20cm) apart and 6 in (15cm) deep. Fill the holes with water to push the seedling roots to the bottom. Leeks like an open position.

harvesting and preparation

Use the seed from the clematis and the leaves of the leek. Gather the clematis seeds when the silky seed heads are mature. Save the outer leaves from the leeks you use as vegetables. Hang the leaves in a net bag to dry (see p. 11).

cooking

Cook about 1/2 lb (225g) leek leaves in a standard alkali solution (see p. 104) for 1—2 hours, until the leaves will pull apart. Allow the leaves to cool in the solution, rinse (see p. 105), then cut them into 1-in (2.5-cm) pieces.

beating method

Blend the cooked leek leaves into a fine pulp (see p. 106). Add the clematis seeds through the feeder during the last few seconds of blending.

sheet forming and drying

Form a sheet (see p. 106) and air-dry it on the mold (see p. 103).

variations

An alternative is to bleach the leeks. Tip the cooked, rinsed, and chopped leaves into a bucket, and add a little bleach. Check the results after 30 minutes. If the bleach is not removing color, add some more and leave for 30 minutes more. Continue until you can see the bleach is working. Leave for 24 hours and you will have a silvery pulp; rinse thoroughly before blending.

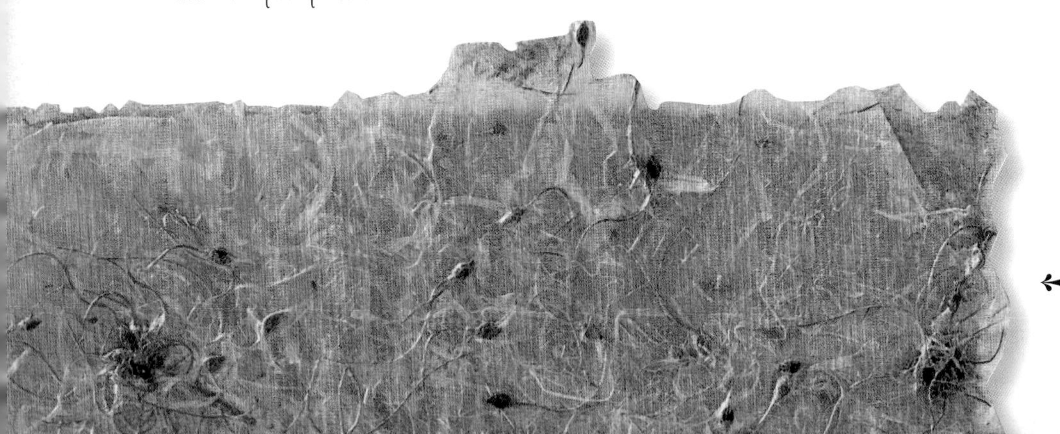

Bleaching the leek pulp gives a paler version

Leek pulp makes a mildly textured paper and the clematis seeds add pattern to it

Mint Tea

Mentha piperita

This paper not only has an interesting peppery appearance, but it also has a lovely minty smell. Peppermint tea is renowned as a cleansing "digestif" drink, so you could prepare a batch of dried leaves and have some for yourself and the rest for making paper.

equipment

Basic papermaking equipment (see pp. 16–18).

cultivation

Peppermint is a native European plant that ideally should be propagated by division or cuttings in spring or fall. Plant in a well-drained soil in sun or partial shade.

harvesting and preparation

Use dried leaves or a commercial peppermint tea. Gather the fresh leaves and dry them (see p. 11), then crumble them into tiny pieces.

cooking

Pour 1 pint (600ml) of boiling water over 2 tsps of peppermint leaves or purchased tea leaves. Leave to cool.

beating method

Add all of the tea, including the leaves to the prepared pulp (see p. 98) and blend briefly to disperse it.

sheet forming and drying

Form a sheet and couch it onto a post (see p. 99–100). Press, then dry by hanging up (see pp. 101–103). This pulp will stain the felts.

Homegrown tea has larger leaf fragments and gives a more densely colored paper

Store-bought teas have very fine pieces, giving an even coloring

Rose-Hip Tea

Rosa

Intermediat

Rose hips are 20 times richer in vitamin C than oranges and have long been used in preserves and sauces. Now they are also popular as a tea. For an interesting extra effect in your paper, sprinkle some extra dry tea over a sheet once it is couched.

equipment

Basic papermaking equipment (see pp. 16—18).

cultivation

None required.

harvesting and preparation

I suggest that you make this paper using a commercially bought rose-hip tea, either loose or in bags.

cooking

Pour 1 pint (600ml) of boiling water over 4 tsps of rose-hip loose tea leaves or the contents of 4 opened rose-hip teabags. Leave to cool.

beating method

Add all of the tea to the prepared pulp (see p. 98) and blend briefly to disperse it.

sheet forming and drying

Form a sheet and couch it onto a post (see pp. 99–100). Press, then dry by hanging up (see pp. 101—103). This pulp will stain the felts. For a rougher texture, air-dry on the mold.

Adding tea to the surface and air-drying on the mold gives a strong stain

This pressed version uses the tea only within the pulp

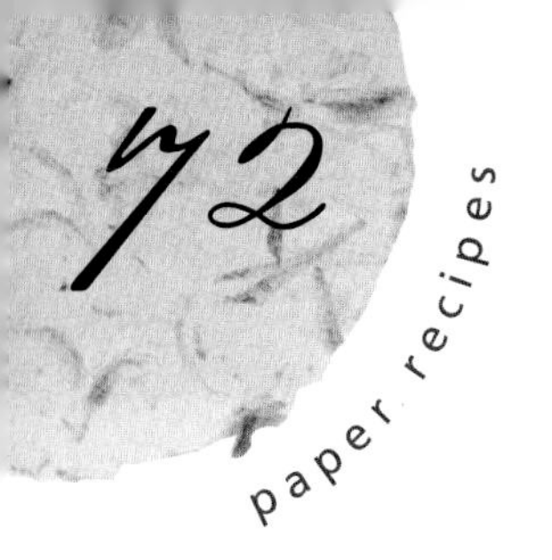

Rooibosch Tea

Aspalathus linearis

The rich tan of the tea tints the pulp while the small stems "bleed" into the sheet as it dries. When air-dried, the stems also contribute to the surface texture of the paper.

equipment

Basic papermaking equipment (see pp. 16–18).

cultivation

Rooibosch (meaning "red bush") only grows on the slopes of the Cedar Mountains in the western region of the Cape Province in South Africa.

harvesting and preparation

Use a commercial brand of rooibosch tea either as loose tea leaves or break open rooibosch teabags.

cooking

Pour 1 pint (600ml) of boiling water over 4 tsps of rooibosch loose tea leaves or the contents of 4 rooibosch teabags. Leave to cool.

beating method

Pour the tea over the prepared pulp (see p. 98) and blend briefly to disperse it.

sheet forming and drying

Form a sheet and couch it onto a post (see pp. 99–100). Press, then dry by hanging up (see pp. 101–103). This pulp will stain the felts.

← *The tea has added both color and texture to the sheet*

Globe Artichoke

Cynara skolymus

At one time the artichoke gained the reputation of being an aphrodisiac, which might explain Catherine de Medici's passion for it. The pure fiber (air-dried) sheet has a very open, free texture.

equipment

Basic and adventurous papermaking equipment (see pp. 16–19).

cultivation

Artichokes need well-drained, rich soil in a sunny yet sheltered spot. Plants have a useful life of three years. They are best bought as plants or cut as offsets from parent plants in spring and raised in soil enriched with plenty of well-rotted manure or compost.

harvesting and preparation

Allow some of the heads to go to seed. Break the seed head open and soak the seeds in water for 24 hours.

cooking

Cook the soaked seeds in the standard alkali solution (see pp. 104–105) for 2 hours. Transfer them to a sieve and rinse under the tap for 3–5 minutes.

beating method

Fill the blender to three-quarters full with water and add a pinch of the treated seeds. Blend for 60 seconds. Transfer the pulp to the vat (see p. 98) and make more batches of pulp in the same way, but alternating the blending time between 60, 30, and 20 seconds, to obtain a mix of fiber lengths. Continue until the vat is sufficiently full.

sheet forming and drying

For a pure version of the paper, form a sheet (see p. 106) and air-dry it on a mold (see p. 103).

variations

For a firmer texture, form a duplex sheet (see p. 107) by couching a sheet of paper pulp onto a post, then couching a sheet of pure seed pulp onto the plain sheet. Press, then dry between newspapers (see p. 107).

The pure pulp paper has a highly fibrous texture

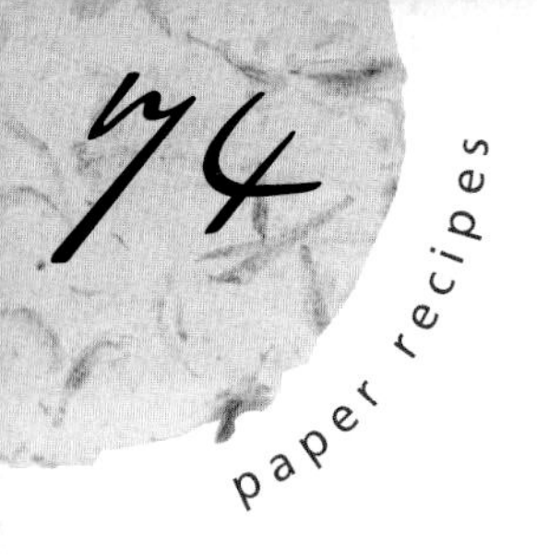

Apple

Malus

Cultivated varieties do not come from seed, and many of the most famous names in the apply dynasty come from random springings from unpromising surroundings — Shaw's Pippin came from a refuse tip and Granny Smiths from an Australian woman's compost heap!

equipment

Basic papermaking and adventurous cooking equipment (see pp. 16—19).

cultivation

Apple trees used to demand a large garden, but modern varieties now take up less room, and some can even be grown in pots. Most apple trees like a good, deep, well-drained loam and an open position on a sun-facing slope.

harvesting and preparation

Use peelings, cores, or whole apples. After you have eaten the best of the crop, use the residue for papermaking. Cut whole apples into quarters.

cooking

Add washing soda to 1 lb (450g) of apples, using 3 tbsps of soda. Add water to cover and simmer for 1 hour, stopping before they become too mushy. Leave to cool, then rinse (see p. 105).

beating method

Quarter fill the blender with the cooked apple and top up to three-quarters full with water. Blend until most of the pieces are very fine. Transfer the mixture to the vat (see p. 98) and repeat the process until the vat is sufficiently full.

sheet forming and drying

Form a sheet (see p. 106) and air-dry it on the mold (see p. 103).

variations

A slightly different character of apple paper can be made by mixing some prepared pulp (see p. 98) with the pure apple pulp in the vat. The result will be a more substantial paper.

← *Apple pulp mixed half-and-half with paper pulp produces a stronger paper*

Pure apple pulp produces a strong, dark color but the paper is fragile →

Coconut Fiber

Intermediate

Cocus nucifera

The deep brown coconut fibers, known as coir, retain their color in the finished paper and their natural dye imparts a harmonious warm tone to the whole sheet. This recipe uses one standard batch of paper pulp (see p. 98) plus about half of the fibers from one coconut.

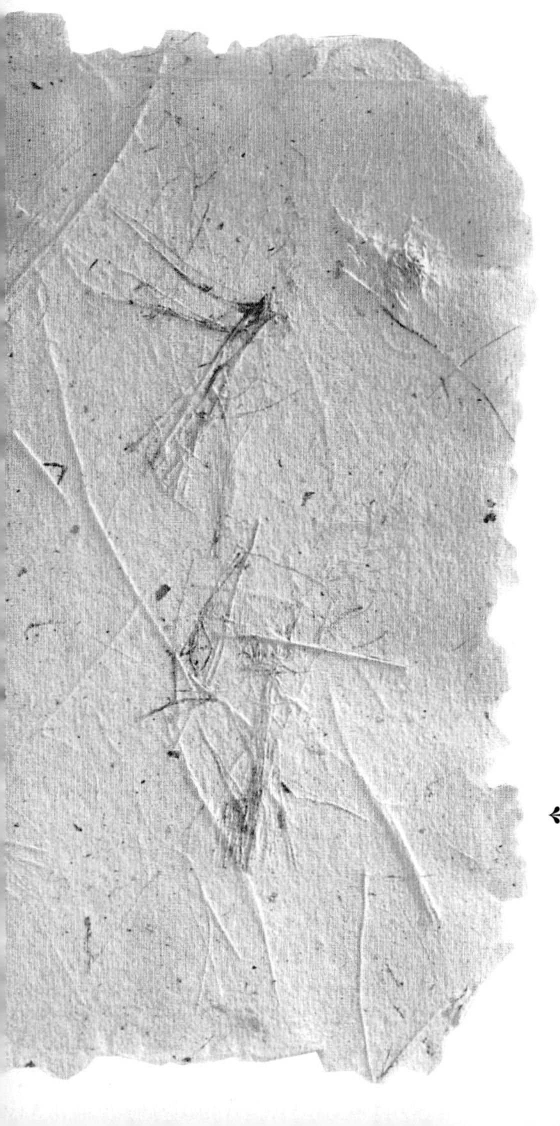

equipment

Basic and intermediate papermaking equipment (see pp. 16–19).

cultivation

The coconut tree only grows in tropical parts of the world, but its fruit is widely exported and can be bought in most good grocery stores.

harvesting and preparation

Peel away the tough brown strands which surround a mature nut. These strands are usually 5 to 6 in (12 to 15cm) long. Cut them into 1/2-in (13-mm) lengths using a sharp knife on a chopping board. Soak the strands in cold water for a day.

cooking

Bring the soaked strands and water to the boil and simmer until the fibers soften and a rich color develops. Leave to cool in the pan.

beating method

Take some of the fibers with some of the cooled liquid and blend them with some of the prepared paper pulp. Deposit that flaskful into the vat and repeat until all the material is mixed. If necessary, add water to obtain the right consistency (see p. 98).

sheet forming and drying

Form a sheet and couch it onto a post (see p. 99–100). Press, then dry by hanging up (see pp. 101–103).

← *A shorter blending time produces a paper with more pronounced fibers*

The soft pink color is created solely by the coconut fiber →

Adventurous

Rhubarb

Rheum rhababarum

This pure pulp paper will appeal for its strong color and lovely texture. Once you have a rhubarb patch established, it is more than likely to produce more than you can eat, so this is a great way of making the most of nature's abundance.

equipment

Basic and adventurous papermaking equipment (see pp. 16–19).

cultivation

Rhubarb is a perennial that thrives in fairly heavy soil to which lots of manure has been added. It can be grown from seed, but it is quicker to buy roots and plant them, either in fall or spring, directly into their permanent position. Even then the plants need a year to establish themselves before the stalks can be pulled in early summer.

harvesting and preparation

Pull, don't cut the stalks. Old rhubarb, which is unsuitable for eating, can be used in papermaking. Cut the stems into 4 to 6-in (10 to 15-cm) lengths.

cooking

Cook 10 stalks in a weak alkali solution (see pp. 104–105) for about 5 minutes, or until the strands pull apart easily. Rinse gently (see p. 105), and test the acidity level (see p. 106).

← *Rhubarb can be food for your table or fodder for your papermaking*

beating method

Fill the blender to three-quarters full with equal amounts of the rhubarb mixture and water. Blend until all lumps are broken up. Do not overbeat since the paper will look best with a mix of particle sizes. Add the mixture to the vat, then repeat, adding extra water until the vat is sufficiently full.

sheet forming and drying

For the pure plant version, form a sheet (see p. 106) and air-dry it on the mold (see p. 103). For a firmer, dulpex sheet, couch a sheet of paper pulp onto a post, then form a sheet of rhubarb pulp and couch it on top of the wet paper sheet (see p. 107). Press, then dry between newspapers (see p. 101–103).

variations

The rinsed fibers, left unbeaten, can be added to a prepared paper pulp (see p. 98) to produce a lighter sheet, beautifully streaked with pink.

Adventurous

Banana Skins

Musa

This fruit is so sustaining that it is said that a man can live for a week on three dozen bananas instead of bread. In this paper the part that usually gets thrown away is used, fresh or dried.

equipment

Basic and adventurous papermaking equipment (see pp. 16—19).

cultivation

Not applicable, unless you are lucky enough to live on a tropical island!

harvesting and preparation

Use commercially-bought bananas for their fruit and save the skins. Dry them in the oven for an hour at 200°F (90°C) or until crisp, since they take too long to dry in a bag, and go moldy.

cooking

Cut 12 to 14 dried skins into 1-in (2.5-cm) lengths and cook in the standard alkali solution (see pp. 104—105) for about 2 hours, or longer, until they pull apart easily. Leave to cool and rinse (see p. 105).

beating method

Fill the blender to three-quarters full with water and add a pinch of banana fiber. Blend until no large pieces remain. Add the mixture to the vat (see p. 100) and repeat until the vat is sufficiently full.

sheet forming and drying

For a pure pulp version, form a sheet and air-dry it on the mold (see pp. 103-106). For a duplex sheet, couch a sheet of paper pulp onto a post, then form a sheet of pure banana pulp and couch it on top of the paper sheet (see p. 107). Press, then dry between newspapers (see p. 101-103).

variations

If you can find them, the leaves of the banana plant, either fresh or dried, can also be included in this recipe. Cut into very small pieces and cook in standard alkali solution for 1 hour. Add the skins and simmer them together for a further 2 hours. Then proceed as above.

The paper made from this waste material is brittle but has a delightful color →

Intermediate

Oakmoss

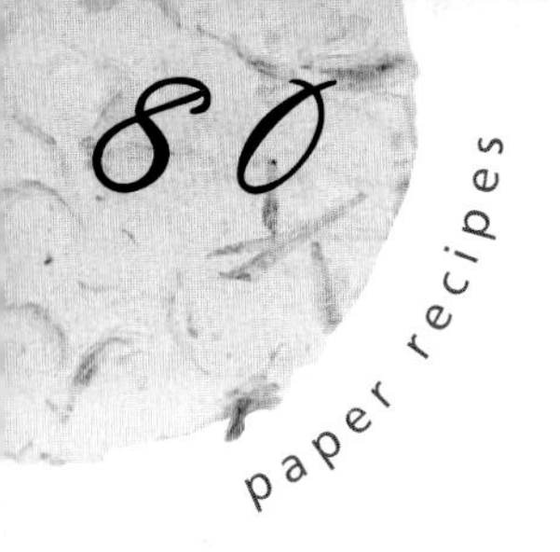

Evernia prunastria

The old apple trees in my garden produce fruit and also give me an additional harvest, the plentiful oakmoss from their trunks. Other mosses can also be used, with a similar effect in the finished paper.

equipment

Basic papermaking equipment (see pp. 16—18).

cultivation

Oakmoss occurs naturally on the bark of old orchard trees. Oakmoss is sometimes erroneously known as lichen, but it is not a true lichen since it is of the Usneaceae genus. Do not use the lichen that you find on rocks — this was once used as a dye plant but it is very slow-growing and is now endangered because it was over-used.

harvesting and preparation

Pick the greeny-blue frills off the tree bark. A handful of moss will make 6 or 7 small sheets.

cooking

Pour 1 pint (600ml) of boiling water over a handful of fronds, which will swell up like little sponges.

beating method

Divide the plant material in two. Lightly blend one batch into the prepared paper pulp (see p. 98). Stir the remainder into the pulp in the vat.

sheet forming and drying

Form a sheet and either couch it onto a post (see pp. 99—100) or air-dry it on the mold (see p. 103). If couching the sheet, press, then dry by hanging up (see p. 103). The air-dried version will have a rougher surface texture.

The very rough texture of this sheet is because it was air-dried

Couched and pressed, the fragments of oakmoss form interesting shapes

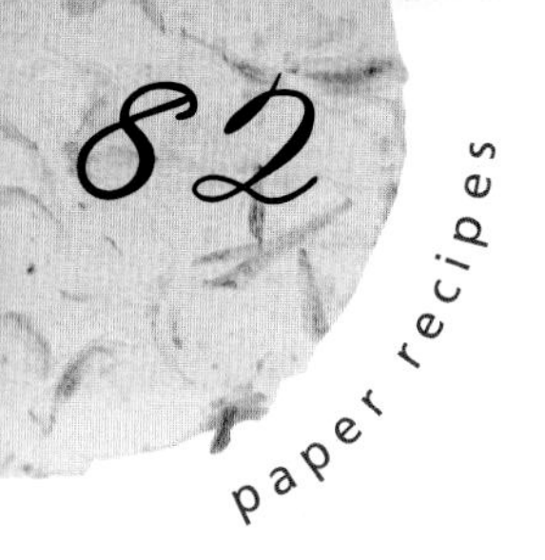

Water Weed

Ranunculus aquatilis and others

This recipe produces a colorful paper which is ideal for using as book covers or endpapers. The curled pondweed Potamogaton crispus can also be used, as can other river weeds and even seaweed. Avoid using the more gelatinous varieties as they can be difficult to rinse and will pucker the paper.

equipment

Basic papermaking equipment (see pp. 16—18).

cultivation

Water crowfoot grows wild in ponds, ditches, and rivers.

harvesting and preparation

Collect the plant in the summer. Half a bucket will be enough for 12—20 sheets, depending on your preferred density of material in the paper. Put the weed on a plastic sheet and discard dead or discolored specimens. Use a sharp kitchen knife on a chopping board to cut it into 1/2-in (13-mm) lengths.

cooking

Pour 2 pints (1.2l) of boiling water over a half bucketful of water crowfoot to get rid of molds and insects. Bring the water back to the boil and simmer for 5 minutes. Pour through a sieve or cloth bag. Make sure no debris such as sand and stones get into the blender.

beating method

Add some of the plant material to the prepared pulp (see p. 98) and blend lightly. Continue adding and blending plant material until you reach the desired density.

sheet forming and drying

Form a sheet and couch it onto a post (see pp. 99—100). Press, then dry hanging up (see pp. 101—103). This pulp will stain the felts.

← Duckweed is another water weed, which gives a different effect

The water weed in this example is water crowfoot, Ranunculus aquatilus →

Thistledown

Adventurous

Cirsium

This soft-textured paper remains quite fragile, but its delicate sheen is particularly beautiful. Thistles are also known as fairies, Susan stealers, or donkey's breakfast.

equipment

Basic and adventurous papermaking equipment (see pp. 16—19).

cultivation

Thistle is a prolific weed, growing freely in the wild.

harvesting and preparation

Use the seeds — or thistledown. When you notice thistledown blowing in the air, in mid- to late-summer, fill paper bags with down pulled from the prickly seed cases. The down will come away easily from fully ripe plants. You will need at least a bucketful, which will reduce to a quarter bucketful once liquid is added.

cooking

Cook 1 bucket of thistledown in half a bucket of standard alkali solution for 3 hours (see pp. 104—105). Let the bucket cool then pour the mixture through a sieve and rinse (see p. 105).

beating method

Take a handful of squeezed thistledown fiber, about the size of an egg, and add to a blender three-quarters full with water. Blend for about 20 seconds, then transfer the mixture to the vat (see p. 98). Repeat until the vat is sufficiently full.

sheet forming and drying

To make a very delicate paper, form a sheet (see p. 106) and leave it to air-dry on the mold. For a stronger paper couch a sheet of paper pulp onto a post (see pp. 100), then form a sheet of thistledown pulp and couch it onto the paper sheet to form a duplex sheet (see p. 107). Press, then dry between newspapers (see pp. 101—103).

The top paper is a duplex sheet, giving it more strength than the pure pulp version underneath

Gossypium

his rope is made from cotton, but other ropes and strings made of natural fibers such as manila, hemp, and jute can be used in the same way, and all produce strongly textured papers.

equipment

asic papermaking equipment (see pp. 16—18) and a coffee grinder.

cultivation

None required, simply recycle old rope hich has outlived its original purpose. ou will need about 1 ounce (30g) of rope or string to make ten 6 x 5 in (15 x 12cm) sheets of paper.

arvesting and preparation

t a length of rope into 1/2-in (13-m) pieces. Put a few dry pieces into a coffee grinder and give it short ursts of power. Keep grinding until the container is full of fluffed-up reads, then empty it out and repeat.

cooking

dd the rope to a bowl of hot water d leave for at least 1 hour until the fibers are fully hydrated.

beating method

None required. Fill the vat with warm water to three-quarters full. Add a handful of soaked rope and stir it in, breaking up any clumps.

sheet forming and drying

When the fibers are as evenly distributed as possible, scoop through the pulp with the mold until the surface is covered (see p. 106). You may have to scoop repeatedly to get an even cover.

Add more fibers and stir before making each subsequent sheet. Air-dry each sheet on the mold (see p. 103).

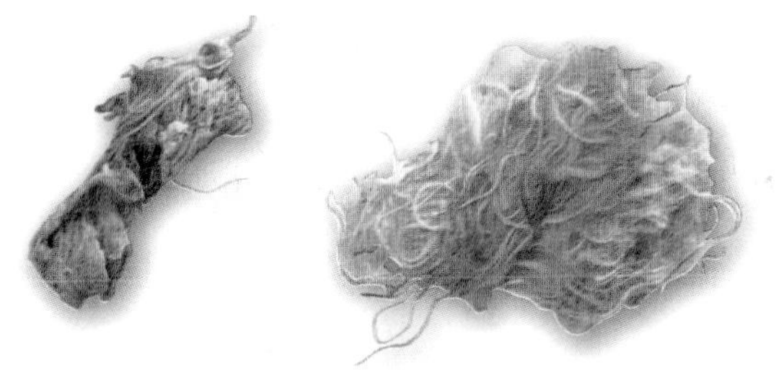

The texture looks strong but the paper is fragile, suitable only for collages

Sisal

Intermedia

Agave

The strands of fiber add a bold texture to the paper and will either appear cream on a white backgroun or pale against a colored pulp. About 1/4 lb (100g) of material is needed for 12 paper sheets.

equipment

Basic papermaking equipment (see pp. 16—18) and a large seashell to use as a scraper.

cultivation

Sisal, or century plant, belongs to the genus of rosetted, perennial succulents and has sword-shaped, sharp-toothed leaves. Sisal needs full sun and well-drained soil. Plants are propagated by offsets in spring or summer.

harvesting and preparation

Collect only a few outer leaves from each plant, to avoid damaging it. Use a large seashell to scrape away the fleshy part of the leaf to release the white inner threads. Rinse off the soft part of the leaf. Cut the fibers into 1/2-in (13-mm) lengths.
Sisal fiber is also available from weaving or craft supply stores, in which case it is ready to use.

cooking

Pour boiling water onto the cut material and leave for at least 1 hou until the sisal is fully hydrated.

beating method

Simply add the sisal to the prepared pulp (see p. 98) and stir it in.

sheet forming and drying

Form a sheet and couch it onto a pos (see pp. 99—100). Press, then dry b hanging up (see pp. 101—103).

variations

To make a translucent sisal sheet, yo can use the Yucca recipe on page 88.

This very pale sheet was made using the yucca recipe (see p. 88)

This dense paper mixes sisal fiber with paper pulp

Yucca

Yucca filamentosa

The yucca, which is actually a member of the lily family, is known by many names: palm lily, Spanish dagger, Adam's needle, or Spanish bayonet. Despite being thin and transparent, this paper is quite strong. It is especially suitable for making lampshades. You will need 1/2 lb (250g) for 12 sheets.

equipment

Basic and advanced papermaking equipment (see pp. 16—19), and a large seashell to use as a scraper.

cultivation

Varieties of this evergreen shrub, an architectural plant with sword-like leaves, vary from frost tender — minimum 7°C (45°F) — to frost hardy. They all need full sun and well-drained soil. Plants are propagated by suckers or division.

harvesting and preparation

Collect only a few outer leaves from each plant, to avoid damaging it. Either use a large seashell to scrape away the soft material surrounding the fibers or pound the leaves with a meat tenderizing mallet. Soak the fibers in cold water for 24 hours.

cooking

Use sodium hydroxide, the strongest of the alkalis, to break down the fibers (see pp. 104—105). Take great care and read the precautions printed on the product and on page 105 before beginning. After completing the alkali treatment, rinse (see p. 105) until the water runs clear and test the pH level (see p. 106).

beating method

Add a small quantity of fibers, about the size of a small egg, to a blender three-quarters full with water. Blend for about 1 minute. Repeat until you have a fine, transparent mixture. This pulp is very thin.

sheet forming and drying

Form the first sheets (see p. 106) and air-dry them on the mold (see p. 103). When you have more confidence with this thin pulp, try couching them onto a post (see p. 100), then press and dry by stacking them between newspapers (see pp. 101—103).

variations

As an alternative to yucca leaves, you could use the leaves of the houseplant Sansevieria to make a similar paper.

The cooked fibers have a translucent quality when made into pure pulp sheets →

Intermediate

Raffia

Raphinus

The alkali treatment on these plant fibers brings out a deep color in the paper. Raffia is readily available and can be bought from weaving or craft supply stores and garden suppliers. Buy the raffia that is described as "undyed natural hanks." Half a hank (about 1 ounce/25g) will make 5 sheets.

equipment

Basic and adventurous papermaking equipment (see pp. 16—19).

cultivation

Raffia is the bast fiber of a tropical climber, so unless you live in the tropics you will be unable to grow this yourself.

harvesting and preparation

Cut it into 1/2-in (13-mm) lengths and soak in cold water for 24 hours.

cooking

Cook for 3 hours in a sodium hydroxide alkali solution (see p. 104). Transfer to a sieve and rinse under the tap for 1 minute (see p. 105).

beating method

Add a small quantity of fibers, about the size of a small egg, to a blender three-quarters full with water. Blend for about 3 bursts of 20 seconds each. Pour the resulting pulp into the vat. Repeat, varying the amount of time you blend each batch, to vary the fiber lengths, until the vat is sufficiently full.

sheet forming and drying

Form a sheet (see p. 106) and air-dry it on the mold (see p. 103) to give a pure plant paper. Alternatively, form a duplex sheet by couching a sheet of paper pulp onto a post and couching a sheet of raffia pulp on top of it (see p. 107). Press, then dry between newspapers (see pp. 101—103).

← *The paper shown here is the air-dried version*

Madder

Rubia tinctorum

Madder, also known as dyer's madder or robbia, has been used as a dye plant since prehistoric times and was of great commercial importance until the end of the 19th century when it was ousted by alizarin, a synthetic substitute. For this paper, however, you must use the real thing.

equipment

Basic and intermediate papermaking equipment (see pp. 16—19), a coffee grinder, and a thermometer.

cultivation

Madder is a hardy perennial that will stand freezing. It can be grown from seed in spring in ordinary soil and the root harvested after three years. Pieces of root with a shoot attached can be replanted in late spring.

harvesting and preparation

Use the root. Harvest the root in late summer and cut into small pieces before reducing further in a coffee grinder. Madder is a dye plant and, as such, the dried root can also be bought from dye suppliers — do not buy the powder form. Reduce the bought root in a coffee grinder.

cooking

Add 2 tbsps of reduced root to 1 pint (600ml) of prepared pulp (see p. 98). Transfer to a saucepan and raise the temperature to 60°C (140°F), testing it with the thermometer. Simmer for 1 hour, but do not let the water heat rise above 70°C (158°F). Take care with the cooking process: if the temperature rises above 70°C (158°F), the color will "sadden" to brown. Remove from the heat and leave to cool.

beating method

Place some of the cooked mixture in a blender filled to three-quarters full with prepared paper pulp. Blend to distribute the color, then transfer to the vat (see p. 98). Repeat the process, stirring to spread the color between batches, until the vat is sufficiently full.

sheet forming and drying

Form a sheet (see pp. 99—100) and air-dry it on the mold (see p. 103).

Paper with madder has the soft, natural color you expect from a traditional dye

Willow Bast

Adventurous

Salix nigra, Salix babylonica, or Salix discolor

Willow trees and shrubs grow wild and are grown for ornament and for basketmaking. Black willow, weeping willow, and pussy willow all produce suitable fibers for this paper. Other bast fibers which can be used include lime (Tilea) and mulberry (Morus), which produces a paper of Japanese character.

equipment

Basic and adventurous papermaking equipment (see pp. 16—19).

cultivation

Grow willow in moderately fertile, moisture-retentive soil in sun or partial shade. Propagate in summer from cuttings. When basketmakers use willow they discard the part used by the papermaker, so a partnership between papermaker and basketmaker would save on unnecessary wastage.

harvesting and preparation

Use the bast, the inner bark. In June or July cut twigs or branches that are not more than 1 in (2.5cm) in diameter and peel the bark from them. Discard the woody center. Soak the bark in cold water overnight then remove the thin, dark, outer bark layer, which should have loosened. If it is stubborn pick it off with tweezers.

cooking

Cook the stripped bark for 2—6 hours in the standard alkali solution (see pp. 104—105). Stop cooking when a trial strand can be easily pulled apart. Let the solution cool and rinse for 5 minutes (see p. 105). The black willow contains a dye and so the rinsing water will not run clear.

beating method

Cut the cooked fiber into 1/2-in (13-mm) lengths and add a small quantity of fiber, about the size of a small egg, to a blender three-quarters full of water. Make batches of pulp, alternating the amount of time each batch is blended for, between 20 and 60 seconds. Add each batch to the vat (see p. 98) until it is sufficiently full.

sheet forming and drying

Form a sheet (see p. 106) and air-dry it on the mold (see p. 103).

Willow bast forms a paper with an open and fibrous texture →

← *From left to right: elm bast, mulberry bast, and bleached lime bast papers. The mulberry has some of the outer bark left on it.*

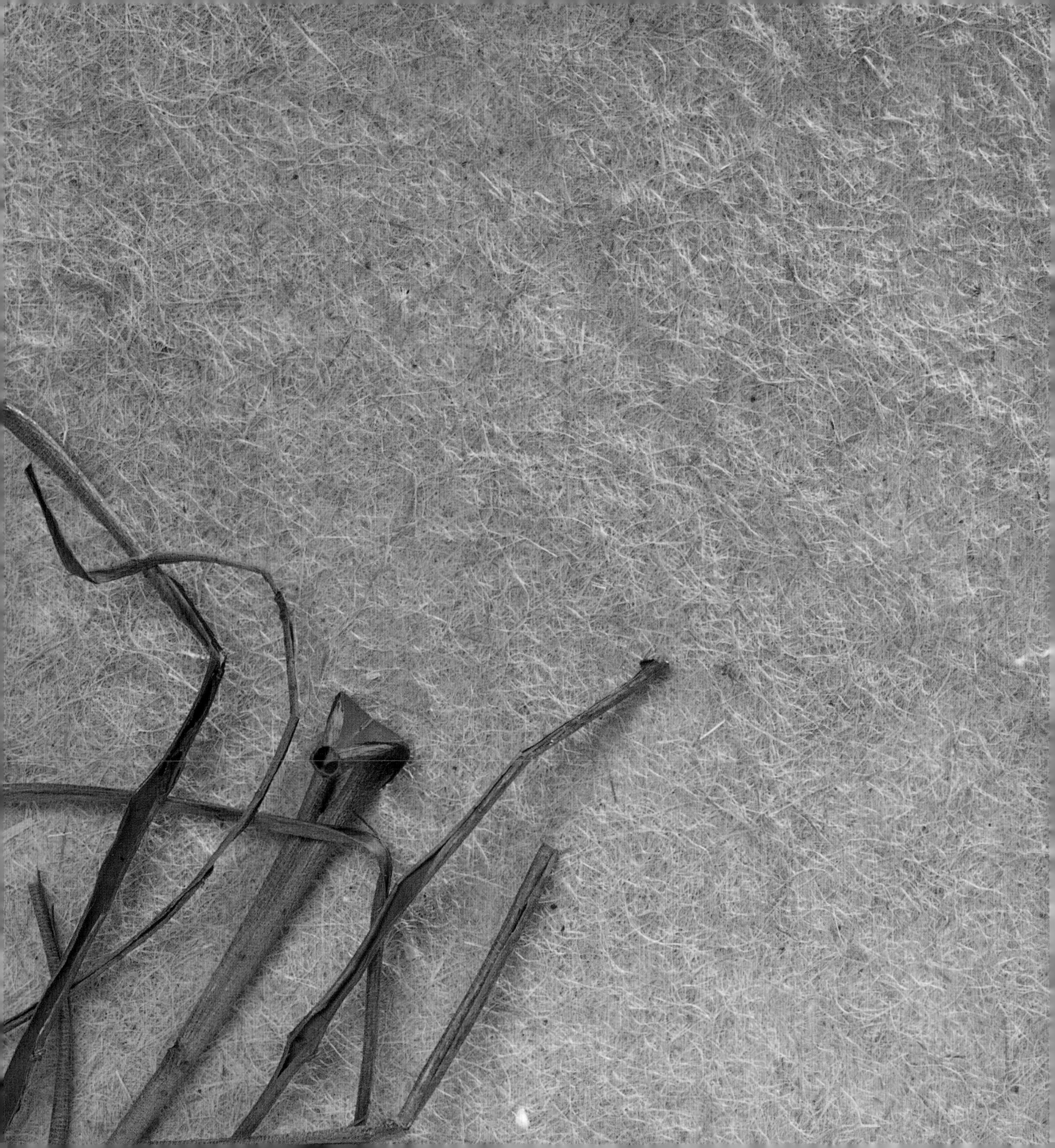

Alkanet

Alkanna tinctoria

Alkanet, also known as anchusa or bugloss, is usually grown for its beautiful blue flowers. By using the root for papermaking you forego the flowers, so I prefer to use the store-bought root.

equipment

Basic and intermediate papermaking equipment (see pp. 16—19), and a coffee grinder.

cultivation

Alkanet is a hardy perennial. It likes a sunny position in well-drained soil. Propagate by root cuttings in winter.

harvesting and preparation

Use the dried root. Either harvest the root in late summer and dry it in a net bag (see p. 11) or buy the dried root from a dye suppliers. Cut it into pieces and reduce further in a coffee grinder.

cooking

Add 2 tbsps of root to 1 pint (600ml) of prepared pulp (see p. 98) and boil until the color develops — up to 2 hours. Leave to cool.

beating method

Place some of the cooked mixture in a blender filled to three-quarters full with prepared paper pulp. Blend to distribute the color, then transfer to the vat (see p. 98). Repeat the process, stirring regularly to spread the color between batches, until the vat is sufficiently full.

sheet forming and drying

Form a sheet (see pp. 99—100) and air-dry it on the mold (see p. 103).

The intensity of color can be varied by using more or less alkanet in the paper pulp

Logwood

Haematoxyin campechianum

Logwood is also known as blackwood or bluewood. If you wish to produce a more intense color of paper, increase the quantity of logwood chips that you use for each batch of paper pulp.

equipment

Basic and intermediate papermaking equipment (see pp. 16—19), and a coffee grinder.

cultivation

Logwood is the heartwood of a fast-growing tree from South America.

harvesting and preparation

Logwood chips can be bought from dye suppliers. Ascertain, from your supplier, that the logwood chips you buy have come from a sustainable source. If the chips are large, reduce them in a coffee grinder.

cooking

Add 4 tbsps of reduced chips to 1 pint (600ml) of prepared pulp and 1 pint (600ml) of cold water in a saucepan. Raise the temperature and, stirring constantly, bring the mixture to just below boiling point. Leave to cool.

This air-dried paper has both strong texture and amazing color →

beating method

Place some of the cooked mixture in a blender filled to three-quarters full with prepared paper pulp. Blend to distribute the color, then transfer to the vat (see p. 98). Repeat the process, stirring regularly to spread the color between batches, until the vat is sufficiently full.

sheet forming and drying

Form a sheet (see pp. 99—100) and air-dry it on the mold (see p. 103).

Papermaking Techniques

A great attraction of hand papermaking is its accessibility. Abundant materials, inexpensive equipment, and modest space requirements are matched by recipes that are as straightforward as an elementary cookbook. The basic technical skills are easily acquired and your abilities will increase with practice. You will achieve ever-more distinguished papers but you will not have to clear any very difficult technical hurdles along the way.

Scooping and couching

First, prepare your pulp with care. Many people are discouraged by their inability to make sheets from pulps with the consistency of lumpy porridge or with unsuitable fibers, such as from newsprint. Make sure your pulp is of a fine consistency, like pancake batter, before you start. Secondly, the smooth action necessary for scooping and couching the wet pulp sheet onto a post of felts is a knack that comes with experience.

Keeping things simple

Fortunately, you can gain confidence by following my experiences. When I first started making paper, I air-dried all my sheets on the mold, avoiding the need to couch and press the papers between felts. I made enough frames to make a number of cockle-free sheets with attractive surface textures. At the very outset, I thought this was a method I had invented myself but a well-traveled visitor, seeing my yard stacked with an array of propped-up frames, remarked that this was the traditional way of making paper in Nepal.

I still use air-drying today, particularly for very fine papers made from pure plants pulps. If failures do occur during your first attempts at papermaking, do not be dismayed. The material is not wasted—put it back in the vat and you will be able to get it right next time!

Simple & Intermediate Techniques

The simple and intermediate recipes all work on a standard quantity of paper pulp, based on eighteen 8 1/2 x 11 in (A4) sheets of wastepaper, mixed with 8 pints (4.5l) of water. This quantity will make 12 new sheets of 6 x 5 in (15 x 12.5cm). Before you start work, cover your surfaces with capillary matting or other water absorbers (see p. 17).

Preparing Recycled Paper Pulp

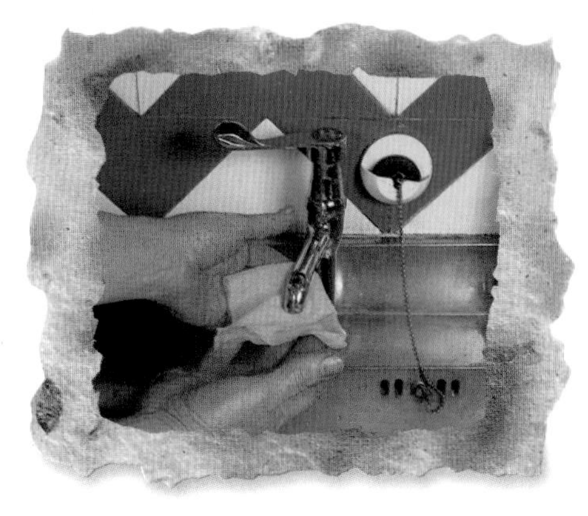

one *To filter the water, wrap several thicknesses of net (see p. 18) around a tap before filling a bucket or kitchen bowl.*

two *Soak eighteen 8 1/2 x 11 in (A4) sheets of wastepaper (see p. 13) in the water overnight, making sure it is completely covered.*

three *Tear the soaked paper into small pieces, about 2 in (5 cm) square. Separate the torn paper into six batches of equal size.*

four *Fill a 2-pint (1.2-l) kitchen blender jar to three-quarters full and add a batch of torn paper. Blend the mixture for 20 seconds or until no undigested pieces of paper remain.*

five *Pour into the vat and repeat. The vat should be three-quarters full. If necessary, add water so the pulp is like pancake batter. Keep the sixth batch aside to top up the vat later.*

six *If you will want to write on the paper, add size to the pulp. Use 2 tbsps of white craft glue (P.V.A.) or laundry starch, mixed with 1/2 pint (300 ml) of warm water. Stir gently.*

Making a Mold

one *A store-bought embroidery frame can be made into a papermaking mold with net curtaining (see pp. 16–17), held taughtly in place. Patterned net will emboss the paper.*

two *Stretcher frame kits, available from art supply stores, can easily be assembled with a small amount of waterproof glue at the joints. Tap into place with a hammer.*

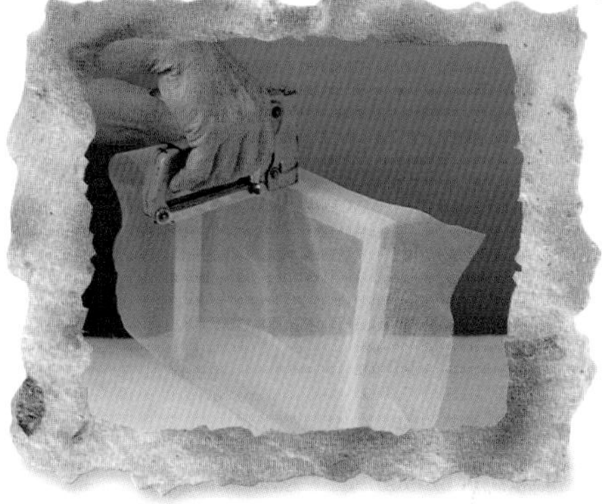

three *Fold net curtaining over the edges and fix in place with stainless steel staples. Align the mesh squarely on the frame and it will be easier to get the tension even all over.*

Sheet Forming

one *Dampen 13 felts, twice as long as the mold and a little wider. Fold one into a small pad, to help release the pulp from the mold. Place another felt so that the pad sits centrally under one half of its length: the other half will be folded over the sheet.*

two *Stir the pulp and wait for the waves to subside. Grasp the mold by the shorter sides and lower it vertically into the vat, at the side furthest from you.*

three *Tilt the top end of the mold downward and pull it toward you until it is horizontal below the surface of the pulp.*

Couching

four *Still holding the mold horizontally, raise it out of the vat. It will emerge covered with a layer of pulp, while water drains back into the vat through the net. Try to complete this scoop in one continuous motion.*

five *Examine the sheet for unevenness. If the sheet is uneven, turn the mold over and gently place it back on the surface of the mixture. Raise the mold and the pulp will drop back into the vat. Stir and scoop again.*

one *When you have an even covering of pulp, hold the mold over the vat for a few seconds to drain. Then position the mold over the couching pad at one end of the felt, turn it over, lower, and press down on the post.*

three *Fold the other half of the felt over the sheet, making a sandwich. Lay a fresh felt on top of the post and repeat until you have completed four sheets. Refill the vat with half the reserve pulp. Make four more sheets and refill the vat with the remaining pulp.*

two *Press a damp sponge into the exposed net to remove more water. Rock the mold from side to side to loosen the sheet before lifting up the mold. After the first few sheets you will be able couch in one continuous motion.*

four *Before completing the final four sheets, allow the pulp to settle and skim over the top with a jug to remove the thin pulp. Sieve the contents of the jug, and return the solids in the sieve to the vat. Stir, let settle, then form and couch the final four sheets.*

Storing Pulp

Sieve the pulp and leave to dry. To use it, soak for two hours and re-blend with water. Wet pulp can be kept in refrigerated airtight containers for a week, longer in a freezer.

Adding Dried Plant Material

In some recipes plant material is positioned on the surface of a sheet. Form and couch the first sheet. Use tweezers to place the plant material as you wish. Fold over the felt.

Pressing

one *Once a post—a pile of sheets and felts—is built, it must be pressed to expel water and help the fibers bond. Cut two pieces of capillary matting (see p. 17) and two boards slightly larger than the felts.*

two *Place the pieces of capillary matting at the base and top of the post and place the post between the boards. Put the post into a press (see p. 18) and tighten as much as possible, until no more water is squeezed out.*

three *To press sheets with bulky additions, or that stain the felts, form and couch one sheet and place the single sheet-and-felt sandwich between several newspapers. Press, and dry by the newspaper method (see p. 102).*

Drying & Finishing

All handmade paper needs to be restrained while it dries. If the pressed sheets were to be removed from their felts while still damp, and left to dry naturally, they would shrink unevenly and cockle. Different methods of drying are appropriate to different types of paper. You cannot use the board drying method for papers with bulky ingredients, but newspaper drying is perfectly suited to them. Air drying method is used for delicate papers that are difficult to couch.

Hanging

indoor *Pick up each felt by the open ends and hang them vertically, with clothes pins (pegs), on a washing line or clothes drying rack. Leave until the sheets are nearly dry (the exact time will depend on how warm and well ventilated the place is) and then finish off by ironing.*

outdoor *To make sheets with an "elephant hide" appearance, hang the felts outside in a breezy spot. Only do this when rain is not forecast, or invest in a rain alert, a device designed to let blind people know when it starts to rain. Once dry, do not iron.*

Newspaper Drying

Place two felt-and-sheet sandwiches side by side on a whole newspaper, not just a single sheet. Cover the felts with at least six sheets of newspaper and position the next two felts on top. Continue to build up a post, finishing with a final whole newspaper. Leave for several hours. Either repeat the process with dry newspaper until the sheets are completely dry, or finish the drying by ironing.

← A sheet of paper that was dried by hanging outdoors, giving a characteristic "elephant hide" appearance.

Photographic Print Dryer

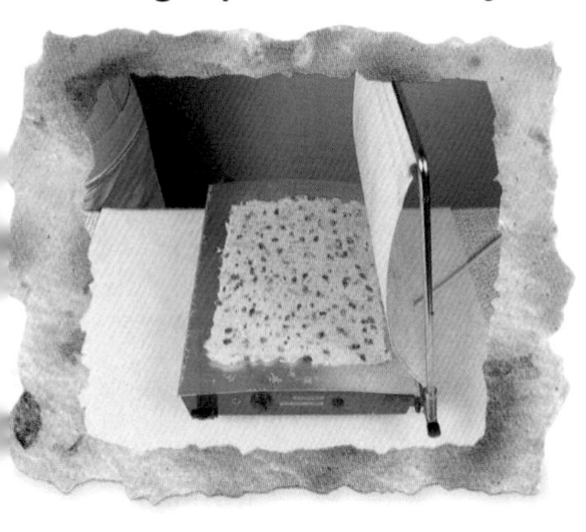

This is great for delicate paper. Partially dry the sheet between newspapers, then lay it face up on the metal plate. Clip the cover down. Use only a low setting; your sheet will be dry in two hours. It will not need ironing.

Board Drying

With fine pulps this method gives you a perfectly smooth sheet. Use a melamine-face board, dampened with water. Unfold the felt and place it sheet down on the board. Brush the felt with a paint-brush, working from the center outward. Remove the felt and leave the exposed sheet to dry for at least 24 hours, undisturbed and away from direct sunlight. If you notice the edges curling, gently spray-mist the sheet with water and sponge the edges back against the board. Once dry, release the sheet with a pointed palette knife.

Air Drying

one *Form a sheet and wipe the pulp off the frame. Let it drain flat until it is only damp. Place it on dry news-paper at an angle so the underside can dry. Finally, stand upright in an airy place, not in direct sun.*

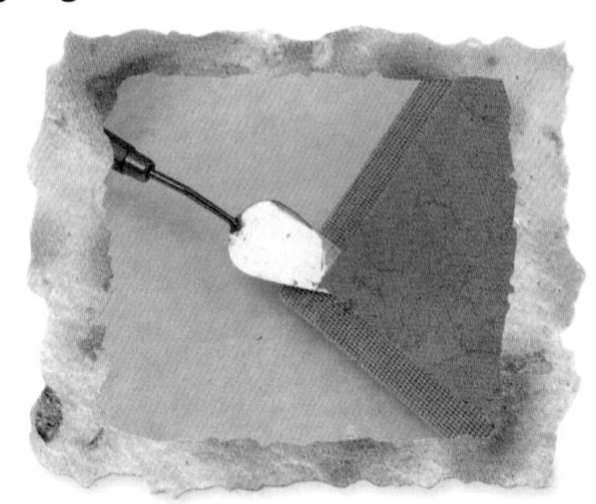

two *When the sheet is completely dry, carefully insert a pointed palette knife where the sheet looks thickest. Support the knife from under the net. Work the blade around the sheet's edges and carefully lift it off.*

Ironing

Ironing finishes a sheet of paper and can help the final stages of drying. Sandwich sheets of dry or semi-dry paper between two sheets of brown paper. Work with a domestic iron on "dry" setting, keeping it moving until the sheet is perfectly dry. Use a very moderate heat for papers with flowers, or it will affect their colors. For a really smooth surface, sprinkle sheets with talcum powder before ironing. Alternatively, mist the sheets with spray starch before ironing: this technique will also help size the paper *(see p. 98)**.*

Adventurous Techniques

The adventurous recipes use pulps made from plant fibers. Treatment of the material with alkaline chemicals releases the cellulose and softens the non-fibrous parts of the plant, which are rinsed away. A standard alkali solution made from wood ash, soda ash, or washing soda is adequate to treat most plant materials. For tougher materials (yucca and raffia) you may need a stronger alkali, caustic soda, but use the weakest alkali possible according to the recipe instructions. Never combine bleach with an alkali solution. If you do want to bleach cooked materials, then make sure the mixture is thoroughly rinsed before you do so. For all recipes that involve cooking with alkali you must wear stout waterproof gloves, a face mask, and goggles.

Preparing the Standard Alkali Solution

one *Lye, made from wood ash (potassium carbonate) makes a standard alkali solution. Perforate the base of a plastic bucket and line it with eight layers of plastic net. Place it inside another bucket to catch the drained liquid. Pierce two sharp sticks through the top of a net bag and balance them on the rim of the bucket. Pour 3 lbs (1.35kg) dry weight of wood ash into the net bag.*

two *Pour 12 pints (6.8l) of boiling water over the ash—you do not have to pour all the water in at the same time but can continue to refill and reboil the kettle.*

three *When all the water has drained through and cooled down, take a pH reading from the liquid in the lower bucket. It should be somewhere between 9 and 12.*

Tip

If wood ash is not readily available, you can make alkali solutions from soda ash (sodium carbonate) or washing soda. Soda ash is available from papermaking and ceramic suppliers. Dissolve it in cold water in a ratio of 2 tbsps of soda ash to 2 pints (1.1l) of water, to give a pH reading between 9 and 11. A convenient, but less pure, alternative to soda ash is washing soda, sold as a household cleaner: here the sodium carbonate is mixed with other, unknown, detergents. Wearing protective gloves, mix the crystals with cold water working to a ratio of 2 tbsps of washing soda to 2 pints (1.2l) of cold water. Mop up any wayward splashes immediately. Different brands of washing soda will produce different pH readings, between 9 and 11.

Preparing a Caustic Soda Solution

Add 2 tbsps of caustic soda to 2 pints (1.1l) cold water (not the other way around as the mixture will froth over), and stir with a wooden spoon until dissolved. It will give a pH reading of between 12 and 14.

Cooking with Alkali

one *Soak the plant material overnight in cold tap water. Strain and cut the material into 2-in (5-cm) lengths. Crush any really hard stalks with a mallet. Transfer the material to your cooking vessels.*

two *Pour the alkali solution over the plant material and add cold water, if necessary, to cover the material. Stir thoroughly, turn on the heat, and keep stirring until the mixture comes to the boil. Do not leave it unattended.*

three *Simmer until the material feels soft. To test it, pull the pieces apart with gloved fingers—they should separate easily.*

four *The cooked material must be well rinsed to remove the non-fibrous particles. Most of the alkali solution will have been neutralized in the cooking process, and so it is safe to rinse it away down the sink.*

five *Collect the cooked material in a sieve (or a net bag if you have bulky material) and test the pH level (see p. 106). If it is acidic, that is pH 7 or lower, then re-treat with more alkali. Rinse under a running tap for 3 to 5 minutes.*

PH Testing

You must check the acidity of the pulp before you proceed to making sheets. Pulp which is too far from neutral—below pH 7 or above pH 8.5—will cause the paper to discolor gradually over time.

Preparing Plant Pulp

Prepare the plant pulp in the same way as preparing recycled paper pulp (see p. 98). Each recipe tells you how much plant material you should add to your three-quarter full blender jar of water. To produce a varied pulp, blend some mixtures longer than others, between 60, 20, and 10 seconds is ideal. This will achieve a matrix of fine, medium, and coarse fibers. If necessary, add more water to the vat to give you enough depth to form the sheets. Don't be deceived by the thinness of the plant pulp—although it is thinner than recycled paper pulp, it will still form a sheet.

Sheet Forming

Scoop the mold through the pulp in the same way as scooping recycled paper pulp and return the pulp to the vat if the layer is uneven (see pp. 99–100). The layer must be especially even if you are planning on air drying your sheets (see p. 103) since areas of different thicknesses will dry at different times, causing the sheet to crack.

Tip

If you find your plant pulp is not producing satisfactory sheets, do not give up. Try adding a little recycled paper pulp to thicken it up. The extra "body" makes the sheets possible to couch, or you can air dry them on the mold.

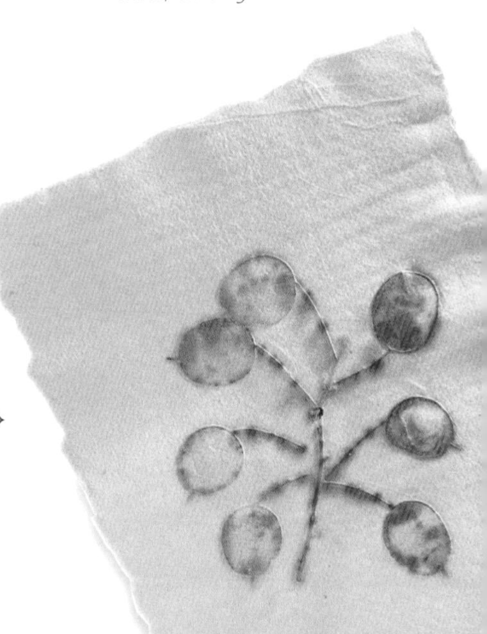

This laminated sheet has a sprig of honesty seed pods between two sheets of plain paper pulp.

Additional Techniques

Once you have mastered the simple techniques of papermaking, and the more adventurous recipes using pure plant pulps, your papermaking can progress to a new level of artstic creativity. The techniques which follow here will allow you to create papers that are personal to your style or customized for particular uses.

Collaging

one *You can build up a collage with wet pulps using the duplex method (see right) so that shapes of different pulps are arranged onto a backing sheet of damp paper pulp. Cut plastic canvas (see p. 17) to the shape you want and scoop it through the pulp.*

two *Couch this over your plain sheet, overlapping with areas of other pulps as desired. Other decorative inclusions can be added at this stage too. Fold the felt over, press between newspapers, then dry between newspapers or in a photographic print dryer (see pp. 101–103).*

Laminating

You can build up three layers of material in one sheet of paper, sandwiching plant material in between two sheets of paper. Form a sheet of recycled paper pulp and couch it. Wet the plant material by soaking it in water or pulling it through the vat of pulp, then place it in position on the sheet. Couch another sheet over the top of the material. Fold the felt over the sheet, press, and dry.

Forming a Duplex Sheet

This is a technique that gives more substance to a sheet and enables those plant pulps that are usually air dried to be couched instead. First form a sheet of recycled paper pulp and couch it (see pp. 99–100). Then form your plant fiber sheet on a clean mold (see p. 106) and couch it on to the damp paper sheet. Fold the felt over the duplex sheet, press, and dry (see pp. 101 and 102). The two sheets will bond together. Duplex sheets can be made using any two pulps and the technique is excellent for making strong, thick sheets: two thin sheets bonded together are stronger than one sheet made with a thick pulp.

This example shows how a wooden printing block (orignally used, in this case, for textiles) can also be used to emboss paper. Take a pressed sheet that is still damp (see p. 101). Unfold half of the felt and lay the sheet face down on the printing block. Press gently with a sponge through the layer of felt. Carefully remove the felt and leave the paper to dry on the block.

Embossing

It is very easy to make patterned imprints on a sheet of paper. One option uses a leaf. Form a sheet of recycled paper pulp and couch it (see pp. 99–100). Place a strongly patterned leaf on the surface of the sheet. Fold over the felt, press, and dry each sheet between newspapers (see pp. 101 and 102). When dry, use tweezers to lift off the leaf.

Watermarks

To personalize your paper, use pliers to fashion a design in fine wire and sew it onto the top surface of the net, with short stitches of linen thread. Form, couch, press, and dry the sheets as usual (see pp. 99–102). The pulp will be thinner over the design, which will show up when the sheet is held up to the light. Remember, the design will be reversed.

Forming Large Sheets

one *Starting small is the best advice for a beginner, but as you become more experienced, and if you have space for larger vats, molds, and presses, you can widen your scope and enlarge the size of your sheets. Alternatively, you can overlap small sheets to make one large sheet. First lay a felt pad (see p. 99), then lay a damp felt larger than the final sheet size over it. Form and couch each normal sheet, overlapping the adjacent sheet by about 1/4 in (1cm), moving the felt pad along each time. Lay a fresh felt over the top and dry between newspapers (see p. 102).*

two *Very large sheets can be made outside. Support a beach mat on strips of wood over a dip in the ground lined with a plastic sheet. Pour the pulp (see p. 98) over the mat using a jug or a turkey baster. It is not necessary to complete the sheet in one session, but more pulp can be added over the space of one week. Move the sheet under cover to dry, which may take as long as a week.*

Scenting

Paper sheets which contain scented flowers or leaves unfortunately do not hold their scent for long, but there are other methods of prolonging or adding scent. Add a few drops of pot pourri reviving oil or essential oil to the pulp in the vat, stirring well so that the drops do not float on the surface and mark the paper. A more lasting scent can be obtained by storing dry sheets with herbs or lavender bunches or with absorbent kitchen paper impregnated with essential oils, in a lidded box for a few weeks.

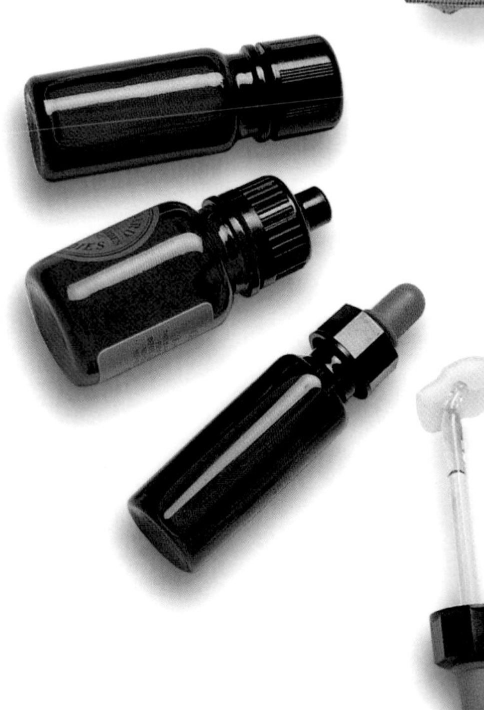

Using a Deckle

Traditionally, paper is made using a mold and deckle (above). The deckle is an empty frame which fits over the mold during scooping and controls the amount of pulp collected. If you want to produce thick sheets that are all of the same weight, or sheets with neat edges, you will need to use a deckle. Although people often refer to paper as having a deckle edge, it in fact gives a plainer edge than just using a mold, which is why I have used a mold throughout these recipes, to emphasize the handmade look of the papers.

Index

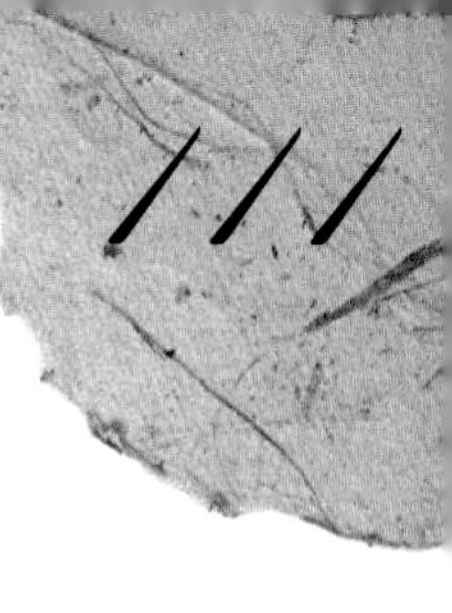

Credits

The author would like to thank:
Brian Richardson, for building the workshop with me, and without whose aid the book would never have been written.
Raphaelle Sadler, who besides working with me on the papers illustrated, helped in many practical ways.
Helen Richardson and *Jaspal Singh*, for making my molds in their picture framing workshop (source of the acid-free mountboard).
Gilly Lewis, *Betty Presland*, and *Shaun Cadman* for contributing flowers, and all my many gardening and papermaking friends and neighbors who gave me advice and support during the research and writing of this book, including an unending supply of dry newspapers.

Quarto Publishing plc would like to thank Random House for permission to reproduce on page 40 the extract from Richard Mabey's *Flora Britannica* (Sinclair-Stevenson, 1996).

For supplies of papermaking equipment and natural dyes, we recommend the following companies:

Carriage House Paper
79 Guernsey Street
Brooklyn
NY 11222
T/F: (718) 599 PULP (7857)
Orders: (800) 669 8781

ColorTrends/Earthues
5129 Ballard Avenue North-West
Seattle
WA 98107
T: (206) 789 1065
e-mail: earthues@aol.com

Twinrocker Papermaking Supplies
P. O. Box 413
Brookston
IN 47923
T: (765) 563 3119
Orders: (800) 757 TWIN (8946)
www.twinrocker.com

Paperwright
1261 Portland Ave
Ottawa, Ontario
K1V 6E8 Canada
T: (613) 731 5417

Earth Guild
33 Haywood Street
Asheville
NC 28801
T: (800) 327 8448
www.earthguild.com

Dharma Trading Company
P. O. Box 150916
San Raphael
CA 94915
T: (800) 542 5227
www.dharmatrading.com